BEST PRACTICES IN
BIOTECHNOLOGY
EDUCATION

BEST PRACTICES IN
BIOTECHNOLOGY EDUCATION

Yali Friedman, Ph.D., Editor

LOGOS
PRESS

BEST PRACTICES IN
BIOTECHNOLOGY EDUCATION
Yali Friedman, Ph.D., Editor

Published in The United States of America
        by
Logos Press, Washington DC
www.Logos-Press.com
info@Logos-Press.com

ISBN: 978-0-9734676-7-3

10 9 8 7 6 5 4 3 2 1

Library of Congress Cataloging-in-Publication Data

Best practices in biotechnology education / Yali Friedman, Editor.
    p. cm.
    "22 International Best Practices in K-12, College, Certificate, Master's, Doctoral, MBA, Distance Education Programs and Student Groups."
    ISBN 978-0-9734676-7-3 (perfect bound)
    1. Biotechnology--Study and teaching--Case studies. I. Friedman, Yali.
    TP248.22.B47 2008
    660.6071--dc22

                        2008005995

Contents

Introduction
Yali Friedman

Biotechnology has impacted diverse fields such as drug development, agriculture, and industrial processes. In addition to the ability to improve human health and quality of life, to reduce our environmental footprint, and even to repair environmental damage, biotechnology also holds the potential to improve economic prosperity. Cities, states, provinces, and countries around the world are aggressively developing strategies and initiatives to develop local biotechnology industries to boost exports, to realize increased tax revenues through more high-paying jobs, and to encourage spillover into and synergies with other high-technology areas.

Developing, attracting, and retaining a workforce to support a local biotechnology industry is a challenge in any location. Rapidly growing biotechnology industries in mature regions are challenged to train sufficient workers to meet local industry demands; developing regions are challenged to train sufficient workers to provide the basis for local investment in biotechnology or to support their fledgling industries. Whether the motivation is to leverage existing resources such as diverse plant and animal species, to commercialize scientific discoveries from an existing research base, or to leverage an existing commercialization infrastructure, the challenge remains the same: biotechnology industries need interdisciplinary workers who understand the diverse factors impacting the biotechnology industry and who can manage the development and growth of biotechnology companies.

The motivation to develop this book emerged from my experiences with *Building Biotechnology*, a book used in biotechnology education programs. In meeting with biotechnology program directors, speaking with students, and providing guidance on program development, I realized that there was an unmet need to collect and share best practices in biotechnology education. Seeking to fill this need, I solicited biotechnology program professors and directors and asked them to describe their programs.

With a solid understanding of the central importance of biotechnology education to workforce and economic development, the authors have provided clear guidance on key issues such as curriculum planning, recruitment, funding, connecting with workforce needs, measuring performance and efficacy, and developing student clubs.

The case studies herein describe a wide variety of programs from high school through Ph.D. programs. Some are in their first years, whereas others are quite mature and have diversified to offer myriad degree and certificate options. There is also strong international representation, with programs from Australia, Canada, New Zealand, South Africa, and the United States.

I hope you enjoy this collection of *Best Practices in Biotechnology Business Education* and look forward to further investigations of this interesting and important topic. I welcome your comments at *info@thinkbiotech.com*.

Yali Friedman, Ph.D.

Biotechnology Entrepreneurship and Education: A South African Experience

Karl Kunert, Juan Vorster, Rachel Chikwamba, and Urte Schlüter

Karl Kunert, Ph.D., is professor in the department of botany and Forestry and Agricultural Biotechnology Institute (FABI) at the University of Pretoria. Karl can be contacted at *karl.kunert@fabi.up.ac.za.*

Juan Vorster is a Ph.D. student at FABI. Juan can be contacted at *jvorster@gmail.com.*

Rachel Chikwamba, Ph.D., is a senior lecturer in the University of Pretoria department of botany and FABI. Rachel can be contacted at *rchikwamba@csir.co.za.*

Urte Schlüter, Ph.D., is a research scientist at FABI. Urte can be contacted at *urte. schluter@fabi.up.ac.za.*

Biotechnology entrepreneurship is new to South Africa, with currently only a very small number of companies.[1,2] However in South Africa, as well as in many other African countries, biotechnology is seen as a future driving force for economic growth in the fields of agriculture, health care, and industry. By developing and enhancing new competencies via education, biotechnology entrepreneurship is considered to promote in particular the establishment of new small, medium and micro-enterprises (SMMEs) thereby significantly contributing to job creation and offering employment opportunities. A successful biotechnology industry therefore has the potential in Africa to make a great impact on both social and economic development.

In South Africa, a successful biotechnology industry might also address some of the historical and socio–economic imbalances. Biotechnology has the great potential to attract young scholars of

1 Louët S. (2006) Rainbow biotech—South Africa's emerging sector. Nature Biotechnology 24:1316.

2 Cloete T.E., Nel C.L., Theron J. (2006) Biotechnology in South Africa. Trends in Biotechnology 24: 557-562.

all races providing the opportunity for (i) skilled job creation, (ii) transfer of skills through international networking, (iii) increased foreign investment, and (iv) the export of high-value products. In addition, a biotechnology industry might address major challenges in Africa by providing products for improved food security through the possible development of genetically modified crops that are acceptable for South Africa and improved health care through the development of vaccines, diagnostics, and treatments for infectious diseases.

BARRIERS TO BIOTECHNOLOGY ENTREPRENEURSHIP IN SOUTH AFRICA

There are currently major constraints on the development of an active and competitive bioentrepreneurship environment in South Africa, hampering short-term success. A first general constraint is the small number of researchers involved in biotechnology who might have potential as a bioentrepreneur. There are fewer researchers in South Africa than in highly technological developed countries. This negatively impacts the availability of sufficient and adequate technical expertise and skills required for start-up companies, particularly in the field of recombinant DNA technology. This skill shortage persists despite the existence of several excellent biotechnology research groups and centers in the country, where research is driven by well-trained and experienced staff. This shortage is also reflected in the relatively low number of patents and scientific publications in biotechnology from South Africa. An encouraging element is that these publications demonstrate the potential capacity to develop a local biotechnology industry in South Africa.

A second constraint is that many new graduates entering the job market lack the skills required to stimulate a biotechnology industry. Further, they rarely have knowledge about biotechnology entrepreneurship and innovation when leaving academic institutions, which maintain their traditional focus on teaching science and executing research. Further, they often have not received extensive training in collection of information required to solve problems, critical review of information, and techniques to search for new solutions. This creates a major problem for companies interested in finding new ideas with potential for commercial success. This lack of skills is also due to both the existing shortage of well-trained faculty in biotechnology in many academic institutions and also the disinterest of many faculty in bioentrepreneurship and commercialization. Since graduates frequently receive only a very narrow specialised training in these institutions, they are further poorly equipped to adapt to rapid changes in technology, which is a necessity for biotechnology.

The limited local opportunities in industrial positions for South African graduates is a third major constraint. South Africa loses many well-trained graduates to developed countries due to a lack of local opportunities. Unfortunately, the current general climate for the development of a general biotechnology industry and bioentrepreneurship in particular is also still not very conducive. A rather complicated funding system, which is less favorable for start-ups, limits the prospects of even the most innovative individuals.

BIOENTREPRENEURSHIP AND SOUTH AFRICAN UNIVERSITIES

There is a great need in South Africa to address these constraints, allowing the development of bioentrepreneurship. Local universities are among the key players. They will have several roles in this process, including education and training, research, and also the provision of infrastructure and facilities for projects with commercial potential in its research phase. They have the fundamental role to provide the required skills and the research outputs in biotechnology that will build the capacity and also the content of those companies that have to be formed to develop commercial products or services in South Africa.

Since universities are multidisciplinary, they are well adapted to the rapid absorption of new technologies and knowledge, a characteristic which has to be strongly supported through sufficient funding and team building. To establish a successful biotechnology industry and support bioentrepreneurship in South Africa, biotechnology graduates should also be encouraged to obtain interdisciplinary training in aspects of law and business, besides traditional training in research, as early as possible. Unfortunately, many students lack the interest to obtain additional skills that are beyond their field of study. Such skills would be integral to a successful career in biotechnology. South African universities should also ultimately shift toward an "entrepreneurial university". An excellent model for this shift is the National University of Singapore. This university is moving toward a knowledge-based strategy for growth to stimulate economic growth through industrially relevant research, technology commercialization, high-tech spin-offs, attracting foreign talent, and inculcating entrepreneurial mindsets among its graduates by injecting a stronger entrepreneurial element in university education.[3]

An additional invisible barrier to entrepreneurship is the scarcity of technology transfer activity within the universities; the individuals graduating from our biotechnology program would be suitable for carrying out such functions and could become agents of change for the academic faculty to be come more interested and involved in science entrepreneurship as a whole.

BIOENTREPRENEURSHIP AT THE UNIVERSITY OF PRETORIA

Although the development of start-up companies in biotechnology is still relatively embryonic in South Africa, there is an increasing trend of start-ups spinning out of universities. We strongly believe that these start-ups will provide a vital source of innovation, ultimately allowing development and commercialization of local biotechnology products and services. However, the current challenges facing the emerging South African biotechnol-

---

3   Wong P-K., Ho Y-P., Singh A. (2007) Towards an "Entrepreneurial University" model to support knowledge-based economic development: The case of the National University of Singapore. World Development 35: 941-958.

ogy set-ups have to be overcome. These challenges include a lack of bioentrepreneurship education in academic institutions, necessary to establish the required mindset for an entrepreneurial culture among graduates.

At the University of Pretoria, the largest university in South Africa, we have addressed this need by recently introducing honours, MSc and PhD degree programs in biotechnology. The most established program is our BSc-Honours Degree in Biotechnology. This program is interdepartmental between the Departments of Biochemistry, Genetics, Microbiology and Plant Pathology and Plant Science. Courses such as *Molecular and Cell Biology (MLB 721)* enable our post-graduate students to carry out biotechnology research within the School of Biological Sciences. The program has traditionally a strong technical component with a major emphasis on gene technologies where our students conduct, under the supervision of a lecturer or senior scientist, a research project as part of the supervisor's research team. A major component within the program is to learn to ask scientific questions, to develop a research strategy for the development of a research product and to finally communicate and defend this strategy in an oral examination. The different departments involved in the program also offer supportive courses by focusing on teaching advanced technologies in biotechnology, general research proposal writing and communication of literature and own research results. However, due to the strong focus on science in the program, we recognized that our students had no adequate exposure to subjects such as bioentrepreneurship and commerce in biotechnology.

Two years ago it was decided that students in our biotechnology honours program not only need to be exposed to science, but also need to be exposed to the aspects of bioentrepreneurship by learning some of the basics required by bioentrepreneurs. With the help of Case Western Reserve University, Cleveland, Ohio, we developed an initial outline for a lecture series called *Biotechnology in the Workplace*, which was part of the *Molecular and Cell Biology* module. For the past two years, all our honours students in biotechnology attended this informal lecture series aimed at addressing these entrepreneurial aspects important for biotechnology start-ups. When we started lecturing and the biotechnology degree was still new, we could only attract a very small number of students interested in a biotechnology honours degree who were also curious to learn something about bioentrepreneurship. Since then, the number has increased significantly and about 40 students have recently shown an interest in our biotechnology honours degree. This increasing interest is also partially due to the fact that younger faculty are becoming more open to the idea of teaching all aspects biotechnology as an important part of university education.

Since the main focus of the *Molecular and Cell Biology* module is to teach our students to carry out research and to develop a research strategy, the entrepreneurship aspects of the course did not develop properly in the last two years in comparison to the science aspects. Therefore, to improve entrepreneurship and also to avoid dilution of the science due to integration of many diverse components into the required course assignments, we decided to introduce a new course called *Biotechnology in the Workplace (BTW 701)* which has its own credits and has to be taken by all biotechnology students in addition to the *Molecular and Cell Biology* module. This course specifically addresses the need for development of bioentrepreneurship in South Africa and is compulsory for our biotechnology honours students. However, since the course is new, we have not yet secured any outside funding. Such funding will be required in the future to invite presentations from local and international entrepreneurs.

BTW 701 focuses solely on introducing our students to topics such as entrepreneurship, intellectual property rights, marketing and financial aspects and how to obtain funding for start-up biotechnology companies. In addition, the new course will also demonstrate the relationship between the academic innovation and its application, with an emphasis on understanding the distinction between academic and industrial research. We secured internal support for the course by including some of the faculty with previous industrial experience as lecturers, and students who were previously involved in the course as mentors, and have since studied different business aspects.

The course schedule, running over two semesters, is designed so our students will develop a research strategy with their supervisors to produce an innovative research product/service which is required in the *Molecular and Cell Biology* module. We ensure that the selected research/service product has the potential to develop into a spin-off biotechnology start-up, and students are required to select an appropriate name for their start-up company. An important component of the course is to connect students with relevant industry issues so they learn more about industrial demands in biotechnology. We therefore invite several local and international bioentrepreneurs with whom our students perform a one-hour interview. The interview includes questions about the personal background and education of the invited entrepreneur, work experience, and what they think are the most important characteristics of a bioentrepreneur. In addition, these entrepreneurs are interviewed about their business, how they started their business, difficulties in writing a business plan and how they obtained funding for their start-up companies. The interview also focuses on team building and the most important characteristics of a successful team in a biotechnology start-up company. A central point is to conduct a market analysis for their products or services and to describe the most challenging problems in setting up the start-up company, the most rewarding aspect, and what type of advice they might give young scientists to set up a start-up company in biotechnology.

A further component of the course is writing a simplified business plan. This component ensures that our students are better trained for developing a start-up company. In the plan they address the seven key areas: (i) the idea or technology used, (ii) the need of the market, (iii) the end market, (iv) intellectual property rights and other issues such as ethical and safety concerns, (v) the team, (vi) possible competition, and (vii) funding. Major emphasis is on selection of a potential new product or service, developing a research product until proof of concept, and forming a start-up business.

Since our students have no business experience, most are very idealistic at the beginning of the course, overestimating the value of their product/service and addressing needs which are often not real. At the end of the course our students are required to present their business plan via an oral presentation. In the presentation, they have to address the key areas and they have finally to defend their business plan before a panel which includes a local bioentrepreneur as an examiner.

CONCLUSION AND FUTURE DIRECTION

We consider our biotechnology entrepreneurship education rather unique for Africa despite programs to promote industrial biotechnology and bioentrepreneurship started by other universities in South Africa. At the University of Pretoria, we also consider the *Biotechnology in the Workplace* module a first step to differentiate post-graduate biotechnology education from the traditional science education offered for them by the different departments within the School of Biological Sciences. Comments from students who were involved in the current lecture series were very positive, highlighting the uniqueness of our approach and requiring inclusion of more business topics in their education in biotechnology.

Through the introduction of *Biotechnology in the Workplace* we believe that we clearly address local South African economic development needs and that our students will ultimately be better equipped for either working in or starting new biotechnology start-ups. Certainly our success will be determined by the number of students enrolling for the course—which is already very encouraging—the demand of our students to be employed in the existing local biotechnology industry, and ultimately by the number of new bioentrepreneurs emerging from our course.

We do not currently expect that our students will start their own business directly after graduating. Our aim is to first create an entrepreneurial spirit, allowing them to seek future career choices toward more entrepreneurial and innovative options, or to proceed with a MSc in biotechnology. The masters thesis will include both a credible science component and a business evaluation of that science. As with the honours course, the masters students will be examined by committees which include both scientists and entrepreneurs who can critically evaluate both aspects of the thesis. Future choices include considering, for example, a future career path in project management, patent law, technology transfer, or business development.

We also think that our course will be a first step for our students to establish a network with the local entrepreneurial community, providing them with energy and commitment to ultimately set up their own companies. For that, we take advantage of the close relationship between the academic domain in Pretoria and the commercial or market-facing domain currently emerging through various local incubator systems. These incubators are in close proximity to the University of Pretoria where faculty and research scientists will increasingly have access to business skills and financial and legal support and also "techno-venture forums" that bring professionals in entrepreneurship and industrial ventures to these incubators to speak.

A future direction to be considered might include the establishment of a separate seed fund specifically for biotechnology students as done by the National University of Singapore.[3] We are also expanding our current course by connecting our students to the local biotechnology industry by offering internships for up to three months, and we will encourage them to participate in an annual national business plan competition. A possible mechanism for supporting these internships would be for the agencies responsible for business development to share the cost of these interns who would be adding value to current small, and often struggling, biotechnology start-ups. Our current course will further be the nucleus for developing advanced bioentrepreneurship education at the University of Pretoria. For scaling up, we intend to offer new elective courses, or specialized workshops and seminars, in new venture creation by specifically targeting MSc and PhD students interested in commercializing an invention.

Finally, from our personal investigations speaking with academics from many African universities, including those in Uganda,

Namibia, Tanzania, and Mozambique, the type of course we are offering, which is flexible due to targeting honours-level beginners with little or no knowledge in the field, might attract considerable interest beyond South Africa. We are currently investigating the possibility of digital multimedia satellite transmission to transmit our course. However, we are aware that certain adjustments to local conditions and also training of local faculty have to be made before any replication of the course can be considered at selected academic institutions in other African countries with an ongoing biotechnology program.

The Biotechnology Certificate Program at Ball State University

Carolyn Vann

Carolyn Vann, Ph.D., is Professor of Biology and Director of the Biotechnology Certificate Program at Ball State University. Carolyn can be contacted at *cvann@bsu.edu*.

Our Biotechnology Certificate Program falls into the category of innovative undergraduate, graduate, and continuing education academic programs. The one- to two-year program developed in 2001 is a certificate program that is independent of a degree that may be simultaneously sought. The program focuses on the technical preparation of individuals desiring to be employed as research associates, seeking employment in forensics, or desiring to gain research experience prior to entering a doctoral program.

The field of biotechnology is one of the world's most rapidly expanding industries. Currently there is a need for individuals who can combine broad theoretical scientific knowledge with excellent laboratory bench skills. In the past and at most universities, science students received a broad background and understanding of science but lacked sufficient practical experiences. Our innovative program is comprised of 10 courses which expose students to current bio-technical approaches and methodologies with primarily lab-based instruction. Six courses are entirely laboratory-based, two others have a very strong laboratory component, and two are traditional supporting electives. Like exceptional athletes, premier scientists must practice their skills. Our goal in this program is to provide students exciting employment or educational opportunities by offering a variety of practical and sophisticated "hands-on" experiences while strengthening their analytical, presentation, communication, and other professional skills.

Acceptance into the program is competitive, with limited enrollment to provide strong mentoring of students by faculty. Currently this program of 23-29 credit hours seeks to attract approximately 16-20 undergraduate/graduate students per year. The hours completed in the program may be applied to an undergraduate or graduate degree—a certificate of completion is provided upon completing the program—and aid in obtaining an internship or job placement is included. Also included in the program is a *Professional Development* class which seeks to permit interactions with researchers in academia and industry and to aid students in presenting themselves professionally. Classes are currently fully enrolled and our graduates report that the program was a very positive and beneficial experience for them. Graduates obtain high paying professional jobs and our placement of graduates is or is close to 100 percent. Numerous employers eagerly seek our graduates and requests exceed the number of trained individuals we can provide.

A full description of the program is at our web site: *www.bsu. edu/biology/biotechnology/*

MOTIVATION FOR THE CREATION OF THE BIOTECHNOLOGY CERTIFICATE PROGRAM

The Biotechnology Certificate Program was developed as a non-traditional, one-year, applied learning experience primarily designed for advanced undergraduates or as part of an MA/MS degree. Shortages of well-trained scientists prepared to address society's rapidly expanding biotechnological needs led to the development of this innovative program. For many years prior to the development of the program our students completing independent research projects were highly sought after because of their exceptional laboratory experiences. In addition, we offered a single practical laboratory-based course, *Research Techniques,* for either graduates or upper level undergraduates. During the previous 10-15 years, over 170 students were placed in biotechnology positions because of the practical training they received. Most of these students completed master's degrees, although some were involved in undergraduate research projects. Following graduation most were employed and remained in Indiana, contributing to a highly skilled technical work force. Our major employers were Eli Lilly, Roche, Dow AgroSciences, The Wells Center for Pediatric Research at Riley Hospital, the Cancer Research Institute, and Midwest Homeostasis & Thrombosis Laboratories.

The new program was designed to provide many more students with additional skills and practical experiences to take advantage of the wealth of opportunities currently available as well as many additional ones anticipated with this expanding technology. At Ball State University we were uniquely positioned to offer this training because of our past experience and because we were concentrating on training at the upper undergraduate/master's level. As highlighted in *Masters of the Biouniverse*,[1] "a traditional master's degree was often little more that a stepping-stone to a Ph.D. or, worse, a booby prize for grad students who failed to earn doctorates." Beginning in the mid 1990s the National Academy of Science called for reform of science education and more relevant preparation of individuals for expanding industry opportunities. Smith examines new professional master's programs that prepare students for jobs in some of the fastest-growing industries in the world—industries such as biotechnology. In 1997 the Alfred P. Sloan Foundation provided the first seed money to launch professional master's programs which has initiated educational reforms beginning to take place across the country. Our Biotechnology Certificate Program was the first in the region to emphasize applied training and provide out of classroom experiences (internships).

---

1   *Smith, F. (2001) Masters of the Biouniverse. HMS Beagle, The BioMedNet Magazine 109*

DEVELOPMENT, FUNDING, AND IMPLEMENTATION OF THE PROGRAM

In the year before the development of the program numerous meetings were held with faculty who might be interested in working in the program, with the departmental chair, the dean, and also with employers and potential employers of our students in order to develop a curriculum and to establish new classes. At this time a life sciences initiative was developing in Indiana so numerous meetings and conferences were attended in order to determine how our program would contribute to the new initiative.

Even though we were relatively late in announcing the initiation of the new program (May 2001), we were able to successfully fill the program (16 students) the first year (primarily our graduating undergraduates desiring to obtain MS degrees) and hired a new contract faculty member to aid in instruction for Fall 2001. Students graduating or completing the program were well received by employers and obtained excellent positions. We also had outstanding employer evaluations of our internship students.

We learned many things from our first year experience.

First, we needed faculty to cover the increased course load. We were given permission to hire a contract faculty member to help with the instruction. The first two contract faculty we hired each worked one year before leaving for outstanding tenure-track positions we helped train them for. We found it necessary to hire a tenure-track faculty member to provide stability. Our other faculty needed some advanced training in newer techniques to develop courses for the program. We were awarded internal funding to develop new laboratories and for training two summers. Although it would be most beneficial to the learning process for students to be exposed to numerous instructors, the new faculty member teaches a majority of the courses while 4-7 others participate in teaching a course or a portion of a course.

Second, graduate assistants were required to help with preparation, setup, and running of the new laboratory-based classes and the chair was reluctant to remove them from other courses. For the first year of the program funding for two assistants was received from an internal grant and from the graduate school. However, we desperately needed graduate student assistance on a continuing basis in this program. In order to support graduate assistants, we initiated an exclusive partnership with InforMax, Inc., maker of Vector NTI software, a very sophisticated package of programs we have integrated into our entire program, including features for DNA sequence mapping, annotation and illustration, PCR primer design, protein and DNA sequence analysis, recombinant cloning strategy design, and many other common bioinformatics tasks. An internal grant was awarded to make the programs available to our students and to place seven copies of the software on our university web server. For a cost of $1000, researchers within Indiana could access our programs as shareware and funds received would support graduate assistant stipends, but also provide equipment/supplies, or be used for student/faculty professional development. A handful of sales were made before Invitrogen bought InforMax and made the software available to all online. Eventually, a few departmental assistantships became available to support the program.

Third, providing an excellent program with the potential for high-paying, exciting, jobs was not sufficient for maintaining enrollment in the program. We needed do a better of informing potential students—especially those not already associated with Ball State—of the program opportunities and of recruiting them. A web site and program brochures were developed. Meetings were held with individuals working in campus admissions and in career services to inform them about our program. In addition, the program was promoted in our undergraduate classes, in local high schools and colleges, and at regional research meetings. Although we saw a decline in applications the second year, enrollments later increased and we have had full enrollment since (~16-20 admitted each year).

Fourth, our program required the purchase of additional sophisticated, large equipment not currently available. NSF equipment grant applications were not funded but we have obtained some second-hand equipment from industry, a LICOR equipment matching grant was awarded us for a sequencer/AFLP analysis/microsatellite analysis system, and our dean has provided us with a new microscopy facility with not only electron microscopes, but also a confocal microscope and teaching/research space. He has also been very helpful in providing us with many new pieces of equipment.

Fifth, the increased emphasis on laboratory exercises required that substantial laboratory supplies be purchased yearly and initially numerous items of moderately priced equipment were needed. We requested that a laboratory fee be charged to cover some of the expenses but this was not approved. An alumni grant was obtained to purchase much of the moderately priced equipment and additional yearly funding from the dean was increased. In addition, our faculty members are excellent about sharing equipment and supplies.

Sixth, the development and continued operation of the program is time consuming. Initially, some summer funding provided support for program development, interactions with industry, recruitment, and advising. However, during the last few years there has not been financial support or course release time for these endeavors. Beginning this fall term, an administrative assistant will take over much of the advising.

OUR CURRENT PROGRAM

Our program at Ball State University is uniquely fitted to fill the niche of providing applied technical training at the advanced undergraduate, graduate, or post-graduate levels. The Program is still unique in the region for providing applied, hands-on, practical laboratory training in small classes, permitting close faculty-student interactions. Students are broadly trained in current laboratory techniques using viruses, bacterial, plants, and animals. They are required to make professional presentations (oral and poster), attend professional meetings, design research projects, write clear laboratory reports and analyses, and to enroll in an internship.

At this time most students in the program are simultaneously seeking master's degrees (primarily MS students). There are always a few students desiring to enroll in only a class or two and who do not enroll in the entire program. Only 3-5 students per year enroll in the certificate program as undergraduates and about half of those complete the program as master's students. About one-third of our master's students completed their undergraduate degree here and another one-third are international students. Most graduate students are supported by teaching assistantships (and a few research awards) which are competitively awarded. The limited number of awards prevents about half of the applicants from attending.

To maintain flexibility in the program, it was developed as a certificate program such that it is independent of the degree being sought. All courses are 400/500 level with enhanced requirements for graduate students. Students completing the certificate program receive a certificate and have the completion of the program designated on their diplomas and transcripts, if they complete the program as either an undergraduate or a graduate student. However, if a student begins the program as an undergraduate and completes it as a graduate student it is not noted on their diploma or transcript, but they do receive a certificate.

We have continued to have full placement of graduates, but more recently they are employed in more diverse positions or enrolled in universities, often outside of Indiana. Although we have continued to place graduates as research associates at Eli Lilly, Roche Diagnostics, Indiana University, and Strand Laboratories in Indiana, numerous students have obtained employment as research associates at Children's Hospital in Cincinnati. Others have found employment at the Medical University of South Carolina, Harvard Medical School, the University of Chicago, and Notre Dame. A number of graduates also have chosen to continue in doctoral programs at Purdue University, the University of Virginia, Penn State University, and Tufts University, to name a few.

CORE COURSES

The following core courses are required for the certificate in addition to two optional courses to be chosen from a list and an internship:

Introduction to Recombinant DNA & RNA Techniques (3 h)
This laboratory course provides fundamental techniques of biotechnology, including experimental design and literature analysis. It is also a survey course, briefly covering many techniques more fully explored in other classes.

Theory and Applications of the Polymerase Chain Reaction (3 h)
Another laboratory course on the design and optimization of PCR, RT-PCR, and real time PCR.

Protein Isolation and Analysis (3 h)
A laboratory course on the theory and application of protein isolation, purification, mutagenesis, relationship of structure to function, and protein-protein interactions.

Professional Development (1 h)
Seminar course emphasizing development of curriculum vitae and discussion of job ethics and values. To maintain connections with industry, seminar speakers are invited to talk with students about careers (some are previous students). Students are professionally interviewed and are provided information on potential internships.

Cell Culture Techniques (2 h)
Theory and practice of cell and tissue culture, including transfections.

Sequence Analysis and Bioinformatics (2 h)
Laboratory exercises and computer database analyses of genomes, proteomics, and evolutionary relationships. Analyses and results

of microarray data obtained from Eli Lilly are presented in poster presentations.

Research Design and Presentation (2 h)
Students design a research proposal using cellular/molecular approaches, make a PowerPoint presentation, and construct a written proposal.

IDEAS FOR THE FUTURE

At one point we wanted to expand into more multi-disciplinary programs that would cross department lines such we're provide some courses for students interested in such careers as criminal justice, patent law, hospital administration, political science. However, working between departments to develop these programs has been difficult as individual departments don't easily share students with other departments and there has not been sufficient administrative support.

We could also provide for continuing education of individuals already working in industry who might need re-training. However, this would require obtaining a facility in Indianapolis convenient to those potential students. In addition, many of our faculty already have other teaching responsibilities in the summer or need this time for their individual research projects and are not inclined to commute. Additional faculty might be needed to accomplish these objectives.

Most recently we have begun changing our previous Ed.D. in Science and Ed.D. in Science Education degree programs into Ph.D. programs. Our Biotechnology classes are expected to a play a very important role in the new doctoral programs, and it is anticipated that we will offer separate graduate and undergraduate courses in the future.

The Colorado Bioentrepreneurship Program Fellowship

Arlen D. Meyers

Arlen D. Meyers, MD, MBA, is Academic Director of the Bioentrepreneurship Program at the Bard Center for Entrepreneurship Development at the University of Colorado at Denver and Health Sciences Center. Arlen can be contacted at *Arlen. meyers@uchsc.edu.*

The $4.3 billion redevelopment of the former Fitzsimons Army Medical Center in Aurora, Colorado into a 578-acre biotechnology park and medical center ignited the Colorado life science industry. By aligning basic life science research with those interested in commercializing the results in an adjacent, emerging bioscience park, the result has been the growth of the Colorado biocluster, which now consists of over 400 medical device, pharmaceutical, biotech and diagnostics companies. The region ranked third in the nation for biotechnology growth.

With the expansion of the industry, it became clear that there would be a need for both technical and managerial talent to fuel its growth.

In March, 2003, the Batelle Memorial Institute was commissioned by members of the emerging Colorado bioscience industry to write a biotechnology strategic plan for Colorado.[1] As a result, they recommended three strategies and 18 action items. The three strategies were:

1. Create a business climate sensitive to, and supportive of, the needs and issues facing bioscience firms
2. Grow the state's biocluster by creating a bioscience

---

1 *http://www.cobioscience.com/stateplan.pdf*

entrepreneurial culture that turns research discoveries into new products, services, and cutting edge firms, and provides appropriate incentives to research institutions and industry

3. Expand the research base and build research excellence in the state's biosciences niche

One of the recommended ways to implement the second strategy was to provide comprehensive in-depth entrepreneurial assistance to bioscience entrepreneurs and companies. Since that publication, several of the proposed action items have been accomplished and progress has been made in building an entrepreneurial infrastructure to assist bioentrepreneurs in Colorado.

In an effort to implement the second recommendation of the Batelle plan, we designed and are now offering a bioscience entrepreneurship training program at the University of Colorado at Denver and Health Sciences (UCHSC) Business School. The program has several key elements including mentoring and support, academic training, networking opportunities, and business development assistance.

OPPORTUNITY

The United Kingdom Department of Trade and Industry defines industry clusters as concentrations of competing, collaborating and interdependent companies and institutions which are connected by a system of market (formal) and non-market (informal) links. Clusters typically include members from four distinct community segments:

- Academic, government, and commercial R&D labs
- Industry
- Public sector support and service agencies, such as regulatory agencies and economic development offices
- Private service sector, such as lawyers, accountants, and consultants

Research suggests that clusters of firms and skilled workers may be one of the key drivers of economic growth in localities, cities and regions. Bioclusters, those clusters concentrating on the development of biotechnology and life science industries, are growing internationally.[2,3,4] In addition, every state in the United States is trying to develop biotechnology as an economic development focus.[5]

The Colorado biocluster, consisting of biotechnology industries, researchers, service providers, and public agencies, now ranks approximately 13th in the nation and is poised for continued growth.[6] As a result, there will be a growing demand for life scientists with business knowledge, skills and attitudes to work in existing firms and new startups.

To take advantage of this opportunity, we have developed a bio-entrepreneurship training program and plan to offer a Bioscience Management Program and an MBA in Bioscience/Biotechnology Program at the University of Colorado at Denver and Health Sciences Center (UCDHSC) Business School to train the future generation of life science leaders, managers and entrepreneurs.

Our target markets are:
1. Graduate level science, technology, engineering, and math (STEM) students interested in commercializing medical devices, pharma/biotech discoveries or health information technology and bioinformatics products.
2. Bioscience and business faculty
3. Mid-level bioscience managers
4. Bioscience service providers
5. Those working in other bioscience interface technologies: aerospace, nanotechnology, energy,

2   http://www.biospectrumindia.com/content/editorial/10511111.asp
3   http://www.nature.com/embor/journal/v7/n2/fig_tab/7400633_f1.html
4   http://www.biospectrumindia.com/content/BioSpecial/103081201.asp
5   http://www.bio.org/local/battelle2006/
6   http://www.cobioscience.com/resources.php

IT, and telecommunications.

6. Business school students interested in bioscience commercialization
7. Bioscience public sector workers e.g. economic development agencies, granting agencies and non-profits

In addition, there is a future opportunity to create combined degree programs in dentistry, graduate science, nursing, and pharmacy, modeled after the existing joint MD/MBA program at CU.

The Colorado Bioentrepreneurship Program (CBP) leverages existing University of Colorado assets including:

- The only graduate level health sciences center training programs in Colorado
- The largest business school in Colorado
- A merger of the HSC and UCD campuses in 2006
- Continued growth of the Colorado biocluster, particularly the Fitzsimons/Anschutz campus
- Existing bioscience programs at the Bard Center for Entrepreneurship Development and the UCDHSC Business School
- The Health Administration Program at the UCDHSC Business School
- The International Business Program at the UCDHSC Business School
- Strong ties to members of the Colorado biocluster

SITUATIONAL ANALYSIS

The Bard Center for Entrepreneurship Development is the graduate level entrepreneurship center at the UCDHSC Business School and is located in the center of the downtown Denver business community. The recently completed Anschutz Medical Campus is located 18 miles to the east in Aurora, Colorado. The

University of Colorado-Boulder campus is located 36 miles to the northwest of Denver.

Bard Center faculty presently offer students 14 graduate level courses in entrepreneurship and a three-course certificate program. In addition, the Bard Center provides incubator space, networking and business development assistance, an entrepreneur-in-residence, venture funds and other support elements for startup entrepreneurs.

In response to the growth of the Colorado biocluster, the consolidation of the UCD campus with the Health Sciences Campus and the perceived demand from students and industry for comprehensive educational programs in bioscience management, the Bard Center created a specialized program in bioscience management and entrepreneurship in 2005. The program was developed by a team representing the interests of the bioscience industry, the public and workforce development sector and academia. The Bard CBP includes networking events, an introductory course in bioscience technology transfer, several more advanced courses in bioscience technology commercialization topics, a special award in the business plan competition and bioscience internships with local biotechnology firms. In addition, an international technology Transfer Fellowship is offered to foreign life science researchers, and a one year technology transfer fellowship for Colorado students started in July 2007.

Present program offerings at the UCD Business School can be seen at *http://www.cudenver.edu/business* and at the Bard Center at *http://www.cudenver.edu/bard* .

In addition to the CBP, the UCDHSC Business School offers a full range of undergraduate and graduate level programs, including nationally respected programs in Health Administration and International Business. A combined MD/MBA program is in place as a joint degree program offered by the business school and the CU Medical School.

THE CBP AT THE BARD CENTER

In response to life science graduate student demands for bio-entrepreneurship education, and based on the success of our International Fellowship Program experience, we now offer a one year bioentrepreneurship fellowship experience for science, technology, engineering, and math/computer science (STEM) graduate students at the University of Colorado. Five fellows who have recently completed their PhD's or postdoctoral experiences were selected using a competitive acceptance process and started the program on July 1, 2007.

The CBP is an intensive year-long program that equips Colorado life scientists with the knowledge, skills, experience, and contacts to start a biobusiness or find employment in the global life science industry. It is anticipated that several of the fellows will remain in Colorado and contribute to the local economy.

CBP offers an integrated series of interactive learning experiences with key players in the Colorado life science field: the University of Colorado Bard Center for Entrepreneurship (CU Bard Center), the Fitzsimons Biobusiness Incubator (FBBi), the University of Colorado Technology Transfer Office (CU TTO), the University of Colorado Denver and Health Sciences Center Alternatives in Science Club (UCDHSC AIS Club) consisting of graduate students looking for academic bioscience career alternatives, the Colorado Bioscience Association (CBSA), and Colorado bioscience entrepreneurs.

The CBP provides a pathway for graduate and postdoctoral life science and engineering students to gain employment in the Colorado and global bioscience industry by delivering academic and real-world technology transfer and bioscience company experience.

The program provides:
- Exposure to the business and technology transfer aspects of bioscience
- An understanding of the legal and regulatory

environment of bioscience commercialization

- Hands-on experience in facilitating and analyzing technology transfer and business development opportunities
- A six-month internship with a Colorado bioscience company—generating an opportunity for a strong employment reference
- Connections to a network of industry, service provider, research, and public sector contacts in the Colorado biocluster

University of Colorado graduate and postdoctoral students with an advanced degree in health, physical, chemical, or mathematical sciences or engineering who are interested in an entrepreneurial career in the Colorado bioscience industry can apply for the program. Of the five fellows who started the program in July 2007, three have recently completed a PhD or post-doc in life science, one is a PhD in the MBA program, and one is a PhD who is working for a national consulting firm.

The program is designed to provide the necessary skills, knowledge, and attitudes required to be successful in either a start-up or existing life science company. In addition to spending time in the CU technology transfer office, the Fitzsimons Biobusiness Incubator and a local life science company, fellows are give scholarships to attend bioentrepreneurship courses at the Bard Center, including a 16-week course in life science technology transfer and four short courses in bioscience finance, new product development, marketing and regulatory affairs and reimbursement.

Successful applicants have a combination of academic expertise in one or more bioscience fields, demonstrated interest and/or experience in innovation/entrepreneurship, and strong personal skills consistent with the potential for future success as a bioentrepreneur. After an initial screening, finalists participate in group interviews with a selection committee composed of academic and industry representatives and entrepreneurs. From this pool, the final group of fellows are chosen.

While the program doesn't guarantee employment, it does provide training in the academic and real-world skills that bioscience companies are looking for in hiring new employees. Through the program, we provide fellows the knowledge, skills and contacts they need to be competitive in the job market.

Table 1: Knowledge, skills, experience and contacts available through the Colorado Bioentrepreneurship program

CONTENT OVERVIEW COLORADO BIOENTREPRENEURSHIP PROGRAM			
	Skills/Knowledge	Experience	Contacts
Technology Transfer Module CU Technology Transfer Office *Months 1-3*	• Analyze scientific results related to commercial drivers and intellectual property (IP) assets • Patent assessment • Produce licensing-oriented marketing tools • Introduction to negotiation and legal document review	Develop business cases involving: • Patentability analysis • Market analysis/ marketing • Patenting an invention • Negotiating a material transfer agreement or other legal agreement • Technology valuation	• CU Bard Center faculty • Potential commercial partners
Technology Transfer Course The Bard Center *Months 3-6*	• Understand the technology transfer process • Learn skills needed to commercialize technology • Evaluate a technology and decide whether to commercialize it or not • Identify technologies that have commercialization potential • Understand how drugs, devices, diagnostics and software are commercialized	Develop and defend an innovation/ commercialization plan based on a life science idea, invention or discovery.	• CU Bard Center and UCDHSC faculty • Experts from Colorado bioscience businesses • Legal, financial, information systems and other industry experts

	Skills/Knowledge	Experience	Contacts
Entrepreneurship Module Fitzsimons Biobusiness Incubator (FBBi) *Months 3-6*	• Scientific and proof-of-concept planning • Patent diligence • Business planning • Market research • Lead clinical indication strategy • Clinical development project planning • Regulatory planning • Financial analysis, modeling and valuation • Prototype development • Manufacture planning • Deal and financing diligence	Develop: • An IP diligence package for a pre-client company • A market positioning & competitive analysis white paper for client • A diligence package for an FBBi company with investor interest Interest areas for these deliverables may include: • Preclinical scientific • Clinical • Financial • Intellectual property • Manufacturing • Regulatory • Marketing	• FBBi Staff and Faculty • Other Colorado bioscience innovators
Courses in Bioentrepreneurship The Bard Center *Months 6-12*	• *Bioscience Marketing* • *Bioscience Finance* • *Managing Regulatory and Reimbursement Processes* • *Managing New Product Development*	Use the core bioscience and business knowledge and skills gained in these courses to complete a capstone project and report	• CU Bard Center business school faculty • Business and bioscience industry specialists
Biointernship Colorado life science company *Months 6-12*	• Overview of bioscience company's strategic priorities and operations	Hands-on learning in a real world setting	• Bioscience company staff • Colorado Bioscience Association members
Learning Community Alternatives in Science Club & Other Venues *Months 1-12*	• Business opportunities in Colorado bioscience • Cutting edge technologies • Strategic issues in Colorado's biotech cluster	Networking with peers and industry experts	• Peers and others with interests in bioscience and entrepreneurship • AIS speakers from the biotech industry • National and local conferences

FUTURE DIRECTIONS AND ISSUES

Since the CBP fellowship has just accepted its first class in 2007, it is too early to know whether the program will accomplish its stated goals. Only using longer-term metrics, e.g. user satisfaction and economic development and technology transfer measures, will answer that question. However, based on our experience with building the bioentrepreneurship at the Bard Center and the CBP Fellowship, we believe there are several critical success factors for such programs.

First, barriers to cooperation need to be eliminated. Since bioentrepreneurship programs, by definition, incorporate and attempt to align diverse academic disciplines in business development, medicine, science, and the law, the barriers to cooperation are significant and must be addressed. Those barriers exist at the departmental, campus, university, and interuniversity level.

Second, proper incentives need to be in place to motivate faculty to participate. Promotion and tenure issues, academic credit for commercialization efforts, salary and equity sharing agreements, and other issues need to be aligned.

Third, since there are serious academic-industry differences, trust and strong relationships between academia and industry partners are pivotal for their success. The programs need to demonstrate that they are addressing a defined market and manpower need and that there is a compelling value proposition to justify the effort.

Fourth, there needs to be a clearly identified respected champion to lead the effort, an interdisciplinary team and industry partners, and adequate administrative support. Program leaders need to clearly articulate goals and benchmarks and need to achieve early successes to sustain the programs.

Fifth, proximity breeds collaboration. The single biggest determinate of biocluster success is the closeness of the research enterprise to business development and technology transfer entities. While geographical distances between research and academic entities can be spanned with modern communications technologies, there is no substitute for face to face interaction. "Face time" and the use of informal networks is another critical success factor.

Finally, program leaders need to be imaginative about funding these programs and building a business model that will guarantee their sustainability.

CONCLUSIONS

We have developed a bioentrepreneurship program at the University of Colorado based on several critical success factors. In the future, we intend to expand the program into a formal MBA in Bioscience Management and define the outcomes. Such programs can substantially contribute to satisfying a market need and contribute to the economic prosperity of the community.

Establishment of a Successful PSM Degree in Biotechnology and Bioinformatics and an Innovative MS Biotechnology and MBA Dual Degree Program

Ching-Hua Wang, Gary A. Berg, and William P. Cordeiro

Ching-Hua Wang, MD, Ph.D., is Professor of Immunology and Microbiology; Chair of Biology, Geology and Nursing programs; and, Director of MS Biotechnology and Bioinformatics programs at California State University Channel Islands. Ching-Hua can be contacted at *ching-hua.wang@csuci.edu*.

Gary A. Berg, Ph.D., is Dean of Extended Education at California State University Channel Islands. Gary can be contacted at *gary.berg@csuci.edu*.

William P. Cordeiro, Ph.D., is Professor of Management and Director of the Martin V. Smith School of Business & Economics at California State University Channel Islands. William can be contacted at *william.cordeiro@csuci.edu*.

Due to the continual expansion and advancement of biotechnology, the need for a well-trained workforce in this vast field requires higher education institutions to rethink their program offerings and to develop innovative and relevant academic programs.

As is well recognized, the United States is at the forefront of the biotechnological revolution with some 1,500 biotechnology companies spread throughout the nation, generating thousands of high paying jobs and billions of dollars in revenue. California leads the nation in the field of biotechnology, having one third of companies residing in the state, many of which are highly successful multinational corporations. Together, California biotech companies employ 260,000 people with estimated revenues of $62 billion.[1]

---

1 *(2006) California's Biomedical Industry Report. California Healthcare Institute and PriceWaterhouseCoopers*

On August 9, 2007, U.S. President George W. Bush signed H.R. 2272 into law, the *America Creating Opportunities to Meaningfully Promote Excellence in Technology, Education and Science Act (COMPETES)*. The final bill includes authorization for the establishment of a Professional Science Master's (PSM) degree clearinghouse at the National Science Foundation (NSF) to share program elements used in successful professional science master's degree programs and other advanced degree programs related to science, technology, engineering and mathematics. It also authorizes the establishment of a grant program at NSF for the purpose of making awards to institutions of higher education to facilitate the creation or improvement of PSM programs.

To meet the needs of the workforce development in biotechnology, the California State University (CSU) system began exploring the concept of professional science masters programs several years ago. California State University Channel Islands (CSUCI) began as a 4-year comprehensive public university in 2001. It is the newest campus in the 23–campus CSU system and is the only comprehensive public university in Ventura County. From inception, the university has been steadfast in its vision to be a forward-looking university serving the needs of the community and beyond. Based on our mission, the faculty and administration of the university have developed and offered a series of innovative academic programs. Two professional master's programs, the MS in Biotechnology and Bioinformatics program and the latest MS Biotechnology and MBA dual degree program, are extraordinarily successful examples of the creation of cutting–edge professional graduate degrees in the sciences.

In this chapter we share our experiences, summarizing the important steps of establishing a professional master's program and describing the essential elements of successful development and implementation of such a degree program. Hopefully, our experiences will help others in their efforts to develop professional master's programs to meet the nation's biotechnology workforce demand.

LAYING THE FOUNDATION FOR PROGRAM DEVELOPMENT

Before initiating a PSM program, the most important step is to have a serious discussion about the idea of professional master's degree programs with all the faculty members in the relevant disciplines. Depending on the nature of the PSM, faculty members could be from one discipline or more disciplines. To lead the discussions, the initiating faculty member should be well-versed in the critical background information regarding the need of the specific PSM program. Relevant information could be obtained by attending regional and national biotechnology conferences, regional economic development meetings, and from discussions with faculty from existing programs elsewhere. Without faculty support and buy-in, it is impossible to successfully launch and sustain a PSM program.

In our case, we only had one faculty member in biology and one responsible for mathematics and computer science in 2001 before the university opened its doors to students in fall of 2002. Due to the limited number of the founding faculty members, these two faculty members placed a MS in Bioinformatics program on our academic master plan. With the growth of our faculty and the concurrent decline of the workforce demand in the field of bioinformatics, we realized that we should not be limiting our degree program to bioinformatics and ignoring the burgeoning field of biotechnology. To learn more about the concept of PSM, the biology program sent a faculty member to a CSU system meeting in the summer of 2003. We also solicited valuable input from one of our sister campuses that had just started a PSM program in biotechnology. We overcame some internal challenges in changing the degree name from MS Bioinformatics to MS in Biotechnology and Bioinformatics and were ready to move forward.

While developing the curriculum for the degree program, we obtained an Extended Education Commission grant for $37,500 from the CSU system to support our curricular development effort. We subsequently conducted a series of surveys: one of the biotechnology industry, one of our students, and one of the CSU system. The surveys were supported by a $6,000 grant from the Sloan Foundation. The data collected from the surveys pointed directly to the need for a professional biotechnology degree program in our region of California. Our surveys indicated that the biotechnology industry needs professionals with strong scientific knowledge and laboratory skills, as well as some foundational training in business management. A key element in our success is having a strong biotechnology industry base in our region; lacking this there would have been little reason for us to develop a PSM in biotechnology.

To understand the biotechnology industry's specific training needs, we visited local biotechnology companies such as Amgen, Baxter and BioSource (now part of Invitrogen) and had panel discussions with the industry leaders, managers and scientists at various levels. We set up an advisory board, consisting of industry leaders, senior management and senior scientists from the local biotech industry, and solicited input from them about the program and from the targeted professionals who would become our students.

We found that bachelor degree-level scientists working in the biotech industry needed advanced training and education for their professional development and career advancement. To retain these junior scientists and attract more into the workforce, companies provide incentives to encourage their employees to attend professional master's programs and other continuing education programs to further their training. In light of housing costs in California, especially in coastal areas, biotech companies increasingly rely on local universities to provide well-trained scientists from local communities for entry level positions. Once employees have worked in the industry for several years, companies often depend on extension programs in regional universities to give them additional training.

CURRICULUM DEVELOPMENT

The outcomes of the discussions with the local biotechnology industry panels and the advisory board members along with the survey data made it clear that we needed to develop a curriculum that should encompass science and business. We also studied the curricula of similar programs in the nation and used them as references in our curriculum development. As a result of our extensive study and research, we first defined our program objectives and student learning outcomes and developed the curriculum accordingly:

Program Objectives:
- Provide students with the opportunity to earn a professional MS degree in biotechnology and Bioinformatics from CSU.
- Prepare students with analytical, business and managerial skills along with sophisticated expertise in biotechnology and computational sciences for a diverse set of vocations. Qualified graduates will be able to engage in research, development and management in biotechnology, work in the pharmaceutical industry or conduct scientific research, teaching or consulting in public and/or private organizations.
- Provide a value added education in biotechnology and bioinformatics to enhance career advancement opportunities.

Student Learning Outcomes:
- Students who successfully complete the Biotechnology Emphasis in the Master of Science Degree program will be able to:
  - Work in cross–disciplinary teams to address questions of relevance to the biotechnology industry through the design and implementation

of databases that integrate computational biology and empirical analyses.

- Explain techniques used to make biological inferences from protein and nucleic acid sequences.
- Identify biologically relevant problems in biotechnology, biomedical, and agricultural research.
- Outline the state and federal regulatory processes that govern the biotechnology industry.
- Explain fundamental principles which underlie modern techniques in biotechnology.
- Demonstrate proficiency in performing fundamental molecular biology techniques.

THE CURRICULUM

COMMON CORE COURSES (16 UNITS):

BINF 500 *DNA and Protein Sequence Analysis (3)*
BIOL 502 *Techniques in Genomics and Proteomics (2)*
BIOL 503 *Biotechnology Law and Regulation (3)*
MGT 471 *Project Management (3)*
BIOL 600 *Team Project (4)*
BIOL 601 *Seminar Series in Biotechnology and Bioinformatics (1)*

FOR BIOTECHNOLOGY EMPHASIS (17 UNITS):

Required Courses (7 units):
BIOL 504 *Molecular Cell Biology (3)*
BIOL 505 *Molecular Structure (4)*
Electives (10 units):
A minimum of 10 units chosen from the following courses and/or from the elective courses under the Bioinformatics Emphasis:
BIOL 506 *Molecular Evolution (4)*

BIOL 507 Pharmacogenomics and Pharmacoproteomics (3)

BIOL 508 Advanced Immunology (4)

BIOL 509 Plant Biotechnology (4)

MGT 421 Human Resource Management (3)

BIOL 490 Special Topics (1-3)

For Bioinformatics Emphasis (18 units):

Required Courses (12 units):

BINF 501 Biological Informatics (3)

BINF 510 Database Systems for Bioinformatics (3)

BINF 511 Computational Genomics (3)

BINF 513 Programming for Bioinformatics (3)

Electives (6 units):

A minimum of two courses chosen from the following and/or from the elective courses under the Biotechnology Emphasis, with at least one course in the BINF category:

BINF 512 Algorithms for Bioinformatics (3)

BINF 514 Statistical Methods in Computational Biology (3)

PHYS 445 Image Analysis and Pattern Recognition (3)

MGT 421 Human Resource Management (3)

BIOL 490 Special Topics (1-3)

The program requires a total of 33 to 34 units to complete. It offers two emphases: biotechnology and bioinformatics. Most of the curriculum consists of courses in science, particularly in molecular biology and computational science. Several key business, law and regulation courses are included as required elements of the curriculum. Courses with 3 units are offered in lecture format. Those with 4 units are lecture and lab combined. Instead of a thesis, students complete a team project, working on a real problem derived from the biotechnology industry. Substantial oral and written reports are required as their culminating experiences in the program. Due to its intensity, this team project course is 4-units. Our faculty members in biology, computer science, business and economics worked together to develop the courses. Each course also has student learning outcomes clearly defined in the course proposals. We developed the curriculum in 2003; the academic approval process took a year; and we were ready to offer the program in Fall 2005. To find the specific learning outcomes for each of the above courses, please visit our website at: *http://biology.csuci.edu*. The website also contains other detailed information regarding our program.

PROMOTING AND LAUNCHING THE PROGRAM

To attract students, we needed to inform the local community about our brand-new program. This was especially important in the first year when we launched the program, fall 2005. We held information sessions on the university campus and in various local biotechnology companies in spring and summer of 2005. We put our program information on our website. The Extended Education Division at CSUCI actively promoted the program in press releases, advertising in local newspapers, sending program brochures and flyers to targeted companies. Since the aims of developing the degree were highly linked to a specific industry, we focused our marketing on current employees in these biotechnology companies. Through purchased industry lists and capitalizing on close industry ties, the recruitment effort was more successful that expected.

Advising roadmaps were developed for students to use for their program of study to facilitate their timely graduation. To meet the special needs of our students who are mostly working adults, our courses were scheduled exclusively during the evenings and weekends, allowing students with full-time jobs to take classes and complete the degree requirements. Student orientations were held to communicate all the special aspects of the degree program to the enrolled students. Industry and advisory board members attended the orientation to meet the students and provide an outlook of the biotechnology industry in the region and beyond. Since we had only three tenure-track biology faculty members with professional expertise limited to their specific fields at the time of program launch, there was no way to offer the program solely with our faculty. With our advisory board members' assistance, we recruited highly qualified senior scientists and business managers as instructors. Appropriately mixing industry professionals with regular faculty for instruction has given the students a multifaceted and real-world based education.

ASSESSMENT OF THE PROGRAM

To assess the program, we require all courses to be evaluated using the same procedure of student evaluation of teaching effectiveness as used in our undergraduate programs. To assess the program effectiveness, we also developed a comprehensive assessment plan for the PSM program that details the assessment activities in relationship to each of the specific student learning outcomes, timelines and how to use the results from the assessment activities to improve our program. The assessment plan also includes the use of a nationally recognized survey instrument (CAEL) that evaluates the specific needs of working adult students.

UNIQUE ASPECTS OF THE PROGRAM

CURRICULUM AND INSTRUCTION

Aside from the MS level courses in the fields of molecular biology and computational biology, we have included several unique courses into our curriculum: *Project Management*, *Biotech Law and Regulation*, *Seminar in Biotechnology and Bioinformatics*, and *Team Project*. The instructors for these unique courses are from the biotechnology industry or working in biotechnology in other capacities. Together, they have many years of experience working as business and project managers, scientific researchers, intellectual property and patent lawyer, venture capital investors and biotechnology company leaders. They represent a cadre of PhD scientists from academia, the biotechnology industry, and managers with strong science backgrounds, MBAs, and an attorney with biotechnology industry experience who specializes in intellectual property law. These instructors offer students relevant examples and on-going trends and challenges in biotechnology. Several science courses are also taught by industry scientists. Again, they have injected biotechnological realism and pragmatism into our program. Our students are very satisfied with the quality of our instructors.

One piece of advice for hiring instructors from industry is to proactively assist them in preparation of their classes. These are highly experienced and highly qualified experts, but they may not be familiar with teaching at the MS level. Some have never taught a class. To maintain quality and meet the expectations of the program, sample syllabi, course materials, and exams should be made available. They should have a clear understanding of the overall program objectives and the student learning outcomes for each course they teach.

ADVISORY BOARD

Another feature of the CSUCI professional master's program is the creation of an external advisory board comprised of industry leaders from throughout the region. Our board has representatives from large, medium, and small biotechnology companies, as well as a community college in the region. Board members offer advice on curriculum, recruitment, internship creation, team projects, and many other issues. They also serve as valuable industry links for students to secure jobs, network, and obtain information about different occupations and fields. We generally meet at least once a semester. The program director reports to the advisors regularly and actively seeks their input on various issues. Our board members are stake-holders of our program and they take a keen interest in encouraging our program's success. They intend to hire our graduates when their companies have employment opportunities. One example of the impact of this advisory board is in the development of the MS Biotechnology & MBA dual degree program, which was initially suggested during an advisory board meeting.

TEAM PROJECT

When we developed our curriculum for the program, our vision was for a final culminating experience for our students based on real biotechnology projects to replace the traditional MS theses derived from research projects. Theoretically, this was an excellent idea. In reality, however, when we were actually implementing the program and were ready to offer the *Team Project* class, we had to abandon the conventional method of curriculum development and be very creative and innovative. We hired Dr. James Harber, director of the Central Coast Biotechnology Center (see elsewhere in this volume), as the Interim Associate Director for our program. Dr. Harber is familiar with the biotechnology industry in our region. The director of the PSM program worked closely with him and organized and coordinated this class for our students.

In spring 2007 we offered the first *Team Project* class. Twenty students enrolled and had the opportunity to work closely with industry advisors from various-sized biotechnology companies. Six teams of students were formed and were each assigned an industry advisor. Among the six teams, subjects varied from reconstructing a human heart with stem cells, application of nanotechnology, building a collaborative molecular biology lab supported by a biotechnology company and CSUCI, improving vaccines for the third world countries, to finding a way to produce cellulosic ethanol from switchgrass. Throughout the semester, students met twice weekly to prepare and organize powerpoint slides. Their out-of-class time was used extensively for research and development of ideas and to gather information for those slides. Students visited the industry advisors' biotechnology companies as well as biotechnology companies working on projects related to their topics. The course organizers, one consultant and one faculty member, set weekly milestones to facilitate a constant flow of work, ideas and communication. Each week the groups reported to the advisors on their collective work, a practice which better prepares the students as communicators, not just scientists.

At the culmination of the semester, a day-long colloquium was held on campus and each team gave an extensive presentation of their work. In addition to the students and faculty, industry advisors as well as university administrators attended the event. The lunch was sponsored by Baxter Healthcare Corporation. The students produced six comprehensive documents and extensive PowerPoint slides summarizing their team projects. The scientists and senior managers included Dr. Ken Feldmann from Ceres, Dr. Tim Osslund from Amgen, Dr. Byeong Chang and Dr. Roger Liu from Biointegrity, Dr. Ken Richards from Invenios, Dr. Bill Tawil from Baxter and Dr. Gil Rishton from the Alzheimer's Institute at CSUCI. These experts spent countless hours working with students on team projects that had real world applications and were impressed with the students' progress.

BUDGETING AND FUNDING THE PROGRAM

To develop the program, we submitted and received two CSU Extended Education Commission grants: one to support the development of the MS Biotechnology and Bioinformatics program and the other to support the development of the MS Biotechnology and MBA dual degree program. These grants helped us conduct surveys, hire consultants, and develop the program proposal and curriculum. The grants also helped us launch the programs.

From the beginning, considering the expertise we had to hire to teach the courses and the equipment and supplies needed to support the program, we recognized that a PSM program like ours would be costly to implement. The faculty collectively decided to propose the program as a self-supported academic program to be offered through our Extended Education Division. Within the CSU, extended education divisions have the authority with proscribed limitations to offer degree programs without the aid of taxpayer funding on a self-support basis. Often such practice is used for innovative new and pilot degree programs. We analyzed and compared the fees charged by similar programs across the nation and at a sister campus. Currently, we charge $795 per credit unit. To complete the Emphasis in Biotechnology, the tuition is

$26,235. Most students are from local biotechnology companies, and these companies often have very generous tuition reimbursement plans. For instance, Amgen offers employees $10,000 per year of tuition reimbursement and Baxter offers $8,000 a year. Consequently, the out-of-pocket expenses for our students are very reasonable.

The Extended Education Division collects all the fees and pays the expenses for instructors, equipment, and supplies. Per a memorandum of understanding agreement between the Extended Education Division and the academic program area, the biology program is reimbursed for the on-going administration and oversight of the professional master's degree. Moreover, the Extended Education Division pays for faculty conducting assessment activities, part of a laboratory technician, consultants as needed, and a student assistant for the program. With this arrangement, our MS Biotechnology and Bioinformatics program has had a balanced budget since the first year of implementation. This is no small feat considering how expensive it is to offer such a technologically intensive program and the select audience! In the last two years, the Biology Program has been able to use the funding for acquisition of additional equipment and supplies to support the PSM program.

LAB RELATED ISSUES AND TECHNICAL SUPPORT

The courses that require laboratory component need technical support. We have recruited several very experienced technicians for our undergraduate program. We use their expertise to assist the instructors in preparation of the laboratory exercises for the PSM program. The Extended Education Division pays a portion of the salary for one technician. In return, the technicians prep the lab classes for the MS level courses. Currently, our lab classes take place in the instructional labs used for the undergraduate program. Soon, with the popularity of the MS Biotechnology and Bioinformatics program, along with the offering of the MS Biotechnology and MBA program, we may have to consider providing additional lab space to accommodate growth of the programs.

FULL SUPPORT FROM THE EXTENDED EDUCATION OFFICE

We receive administrative support in addition to budgetary allocations from the Extended Education Division. The Extended Education Division organizes advisory board meetings, produces program brochures, distributes student recruitment materials, keeps student records, corresponds with applicants, students, and instructors, orders textbooks, administers grants, conducts surveys, and performs graduation checks, among other functions. Without the full and effective support from the Extended Education Division, we would have not been able to implement our program and make it successful.

CULTIVATION OF TEAM SPIRIT AMONG STUDENTS

Almost all students in our program are working adults. Our program recognizes the importance for these students to build a team spirit as one of the desirable outcomes of our program. Therefore we organize several social events to give students an opportunity to interact outside of classes. Through such social activities and related networking, some students have found employment at local biotechnology companies.

EMERGING NEW PROGRAM: MS BIOTECHNOLOGY / MBA DUAL DEGREE PROGRAM

With the success of our MS in Biotechnology and Bioinformatics program, and with the input from our advisory board members, we realized a need for people who are well-trained in sciences and are well-versed in business management at advanced levels. Once again, we conducted national surveys and had extensive faculty discussions among the biology and business and economics programs. The need was evident from our surveys and extensive research. We also searched for similar programs in the nation. Surprisingly, we found only a handful of such dual degree programs, all offered by universities on the east coast. Clearly, there is a need for us to fill.

Since both our MS in Biotechnology and Bioinformatics program and the MBA program have been very successful, enrolling numerous students from our community, we essentially meshed the curricula of the two existing programs together to build the dual degree program.[2] As both the MS Biotechnology and Bioinformatics program and the MBA program were approved by our academic senate and CSU system office previously, we were only required to submit a program modification proposal to our campus curriculum committee. Due to the strong track record we developed for both stand-alone programs, approval for the dual degree program went smoothly. In fall 2007, we are offering this program for the first time and have admitted 41 students with 31 enrolled in the program. By the number of inquiries alone, we expect that this program will definitely attract more interest in future years.

TIMELINE AND GROWTH OF THE PROGRAM

The following is the timeline for our program development and implementation. It shows the important milestones of starting and offering our PSM and the time it took for each step. It might be useful for other programs to use as a reference, especially for colleagues who would like to start a PSM at their universities.

Fall 2002: proposed the PSM program as MS in Bioinformatics to the academic master plan on campus.
2002-2003 Academic Year: discussed and researched the concept and practices of related PSMs among faculty, decided to expand the degree program to be MS Biotechnology and Bioinformatics.
2002-2003: submitted and awarded Extended Education

_____
2   *http://biology.csuci.edu*

Commission grant from the CSU and the Sloan Foundation Grant to support the development of the professional MS in Biotechnology and Bioinformatics program.

Spring 2003: conducted local, regional and state-wide surveys and visited local biotechnology companies.

Spring 2003: set up the Advisory Board for the MS in Biotechnology and Bioinformatics program.

Fall 2003: developed the curriculum for the MS in Biotechnology and Bioinformatics program.

Spring 2004: received the approval of curriculum by the Academic Senate at CSUCI.

Spring 2004: submitted the Program Proposal to the CSU Chancellor's Office for approval.

Spring 2005: received approval from the CSU Chancellor's Office as a pilot program starting fall 2005.

2005-2006: submitted and awarded Extended Education Commission grant from the CSU to develop a dual de-gree program of MS Biotechnology and MBA.

Spring and summer, 2005: recruited students, sent out advertise-ment of PSM program in the community and biotechnol-ogy companies in the region.

Fall 2005: accepted the first cohort of 23 MS students; offered course to them as a cohort.

Fall 2006: developed the curriculum for the MS biotechnology and MBA dual degree program and submitted to the Academic Senate's Curriculum Committee for approval.

Fall 2006: admitted 40 MS students as the second cohort.

Spring 2007: admitted 5 additional MS students.

Spring 2007: received approval from the Academic Senate to of-fer the MS Biotechnology and MBA dual degree pro-gram, starting fall 2007.

Spring 2007: graduated 20 students from the first cohort in the MS biotechnology and Bioinformatics program; Most of the graduates moved on to the MBA program, starting fall 2007.

Spring and summer, 2007: admitted 48 new MS students for fall 2007.

The admission process is on-going. Twelve of the newly admitted students for fall 2007 are seeking the dual degree, in addition to most of the graduates from the first cohort of MS Biotechnology and Bioinformatics program.

Fall 2007 is also the first time we are accepting international students. We have admitted 6 international students, representing 4 countries and received over 250 inquiries regarding our PSM program from overseas students! Due to the ever-expanding student body, we are no longer organizing students as cohorts. They are free to register in classes and progress at their own pace. Additionally, we switched to a four-term 12-week schedule to better mesh with the MBA program and meet the needs of working adults who want accelerated degree programs. To facilitate their progress to degree completion, we provide students in both programs with an advising roadmap for graduation as a guideline. According to our roadmaps, a student enrolled in the MS in Biotechnology and Bioinformatics program could take 2 to 3 courses a quarter while working full-time, and complete the degree in one year. If they take 1 to 2 courses a quarter, they could complete the program in 2 years. For those enrolled in the dual degree program, they could finish their degree in 2 or 2.5 years, if they take 3 to 4 courses or 2 courses a quarter, respectively.

FUTURE PLANS

CSUCI is in the process of developing a research incubator and a research and development park, which will establish mutually beneficial relationships with the local biotechnology industry and our two professional master's programs. As we go forward we are conscious of the need to be extraordinarily flexible and responsive to the changing needs of the local industry and the scientific community. We view plans for a research park and incubator as a way to serve our mission as a university by meeting the evolving needs of local industry, educating a new generation of scientists, and supporting the original research of our faculty.

A Model for Connecting Students and Teachers to the Biotechnology Industry Cluster in San Diego County

Sandra Slivka and Ashley Wildrick

Sandra Slivka, Ph.D., is Director of the Southern California Biotechnology Center @ Miramar College. Sandra can be contacted at *sslivka@sdccd.edu*.

Ashley Wildrick is Special Initiatives Program Manager at the San Diego Workforce Partnership Inc. Ashley can be contacted at *ashley@workforce.org*.

In an effort to prepare a world-class scientific workforce for San Diego, BIOCOM, the industry trade organization and the San Diego Workforce Partnership Inc. have developed a model program that exposes students and teachers to the regions life sciences industry cluster. Ranked third in the world, this industry cluster of over 500 companies and research institutions faces a local workforce shortage as the industry continues to grow and expand. This impending labor shortage led the Department of Labor to fund the San Diego Workforce Partnership Inc. under the Presidents' High Growth Job Training Initiative, as implemented by the US Department of Labor's Employment and Training Administration. The Life Sciences Summer Institute (LSSI) is a collaborative product of this funding to the San Diego Workforce Partnership Inc. and BIOCOM. Initiated in the summer of 2005, the LSSI seeks to foster interest in the life sciences among upper-level high school, community college and university students as well as high school teachers. The ultimate goal of the program is to give San Diego's future workforce early exposure to the life sciences industries.

This model has three specific objectives:

- To provide industry with well-prepared interns
- To provide students hands-on experience within the life sciences industry
- To better equip teachers to prepare our future workforce

This sustainable model addresses both the impending labor shortage as well as the math and science education crisis facing our nation's educational system.

KEY COMPONENTS OF THE PROGRAM

Initial efforts in late 2004 brought industry and education together to develop the program. The key elements were developed initially and several refinements have been made over the past three years based on continuous feedback from key stakeholders, students, teachers, and industry.

STUDENT PROGRAM

- Student Selection Process: A "temp agency" placement model where student applications are pooled and students are interviewed and 'hired' by host institutions
- *Boot Camp: Introduction to the Biotechnology Industry.* One unit college credit class at San Diego Miramar College prior to the start of the industry internship (one week of industry defined hard and soft skills).
- Industry Internship: A 7-10 week industry or research institute experience paid for by host institution
- Student Exhibition: A follow on event "Celebration of Science Education" where student posters are exhibited to the life science community

TEACHER PROGRAM

- Curriculum Training (Standards Based): High school and community college teachers are trained on the Amgen-Bruce Wallace Biotechnology Laboratory Program at the Biogen Idec Community Lab
- Externships: Teachers visit a variety of industry sites for half day 'externships' to view both hard skills and soft skills in practice. Exposed to working professionals *in situ,* these teachers can explore the expectations of educational outcomes in the workplace.
- Curriculum Sharing & Peer Networking: Teachers have the opportunity to share-out "best practices" and network amongst each other.
- Ongoing Support for Curriculum Implementation: Teachers who have no equipment or follow on support receive free supplies, loaner equipment, and staff support to implement the curriculum (Grant funding from Amgen Foundation).

THE PARTNERS

IMPLEMENTATION

- The San Diego Workforce Partnership Inc. has been coordinating job training and employment programs for more than thirty years. Created under a Joint Powers Agreement by the City and County of San Diego, the Workforce Partnership brings qualified employees and area businesses together.
- BIOCOM, the industry and trade association for San Diego's life sciences community. The BIOCOM education and workforce committees serve as steering committees for these programs.

MAJOR SUPPORT: INDUSTRY AND FOUNDATIONS

Biogen Idec has generously hosted the LSSI Teacher Externship Program for the past three years in their state-of-the art Community Lab facility, in addition to hosting student interns and providing teacher externship experiences. One of the unique aspects of the Community Lab is that it takes teachers out of the classroom and brings them into a working environment where science is applied every day.

> *"Biogen Idec, the world's third largest biotechnology company, is extremely interested in increasing the pool of potential scientists that could become our employees, especially from within the communities that have been traditionally underrepresented in science. Our partnership in the Life Science Summer Institute provides Biogen Idec with the opportunity to reach even more students and teachers. Biogen Idec is committed to the continued use of our Community Lab for the duration of the LSSI, providing teachers with access to state-of-the art equipment and biotechnology professionals at all levels of our company."*
> Annie Glidden, Director, Biogen Idec Community Lab

One of San Diego's largest biotechnology companies, Invitrogen, has also partnered with the LSSI programs to advance and promote science education throughout the region. Invitrogen has donated thousands of dollars in Invitrogen products to supplement both the student and teacher laboratory training programs, and has more recently donated $7,500 to support sustainability efforts of the LSSI Teacher Externship Program.

> *"From student internships to teacher externships, we are excited about the LSSI programs and believe they will significantly improve scientific literacy at all age levels. Invitrogen's goal is to inspire students and teachers to continue their pursuit of science, and the LSSI program objectives and outcomes align with our own perfectly. Through hands-on, interactive*

programs and dynamic curriculum, LSSI is a valuable partner in growing California's future workforce."
Lisa Peterson, Community Relations Manager, Invitrogen Corporation

On behalf of the LSSI Teacher Consortium, the Southern California Biotechnology Center (SCBC) applied for and was awarded a three-year grant for the expansion of the Amgen–Bruce Wallace Biotechnology Laboratory Program into San Diego County, through current and future LSSI graduates. The funding obtained allowed for the purchase of equipment and supplies to make transferable "laboratory kits" for rotation throughout county classrooms. An outreach coordinator was hired to oversee the distribution of these kits. The outreach coordinator responsibilities also include working with the teachers throughout the LSSI program, and providing ongoing support to LSSI teachers implementing the Amgen curriculum into their classrooms throughout the school year.

"I feel that we have really made a breakthrough in that we now have a mechanism and process in place to help teachers implement new lab curriculum covered during the LSSI into their classrooms."
Martin Ikkanda, Professor of Biology, Pierce College (LSSI Teacher Curriculum Instructor, Amgen-Bruce Wallace Biotechnology Lab Program)

MAJOR SUPPORT: EDUCATION

- The Southern California Biotechnology Center (SCBC) @ Miramar College, one of six CalABC Centers (California Applied Biotechnology Centers) funded by the California Community College Chancellors Office, works to prepare the local workforce for the regional industry cluster. The SCBC developed the hands-on laboratory curriculum utilized during the student biotechnology "boot camp" trainings at Miramar College, allowing the use of the laboratory equipment and facilities. The biotechnology "boot camp" is an accredited course offered by the college (*Introduction to the Biotechnology Lab*). Each student who successfully completes the course receives one full unit of college credit. In addition to creating the training curriculum, and providing laboratory space and equipment, the instructor fees are also paid for by the college.

- California State University, San Marcos, provides the option for teachers to obtain credit for their participation in the program. The units earned by the teachers are academic semester units that are reported on official university transcripts with a grade attached.

HISTORY OF THE LSSI

CATALYST FOR CREATION OF THE PROGRAM

A President's High Growth Job Training Initiative Grant obtained by the San Diego Workforce Partnership Inc. from the U.S. Department of Labor, Employment and Training Administration, and with support and leadership from Biogen Idec and the Southern California Biotechnology Center (SCBC) at Miramar College, the Life Sciences Summer Institute (LSSI) kicked off in the summer of 2005. This funding fully supported the program for the first three years. A model of sustainability in the absence of such grant funding is one of the current challenges facing the program.

DEVELOPMENT OF THE PROGRAM: AN EMBEDDED CONNECTION TO BOTH INDUSTRY AND EDUCATION PROVIDES INDUSTRY RELEVANT EXPERIENCE

The San Diego Workforce Partnership Inc., along with its partner BIOCOM established a taskforce committee, comprised of industry professionals (community relations and human resources representatives, hiring managers, and scientists) and educators (San Diego County Office of Education K-12 science coordinators, community college and university science department chairs, internship coordinators, and school administrators) to formulate and design the internship and externship programs for students and teachers. This cross-functional and diverse taskforce conducted a needs assessment of the region prior to breaking off into sub-groups to further work on program design, curriculum development and outreach models for the initiatives identified.

In addition to this Taskforce and working sub-groups, BIOCOM facilitates several committees to ensure the needs of the industry are being served and partnerships are formed. Members from life sciences companies, research institutes, academic institutions, staffing agencies, the San Diego Workforce Partnership, and the State of California Employment Development Department (EDD) meet regularly to discuss workforce needs, the educational pipeline and career ladders. At the board level, presidents and CEOs steer initiatives through the Workforce Capabilities Committee. Under this committee are the human resources subcommittee and education subcommittee that provide feedback and make decisions and recommendations for the further development and continued success of the LSSI programs. This structure allows for maximum input from interested parties, and fosters participation from industry at multiple levels.

INDUSTRY RELEVANT CURRICULUM & TRAINING

STUDENTS: BOOT CAMP TRAINING & INTERNSHIP

This one week intensive lab training with soft skill awareness exposes the students to a working biology laboratory as it relates to the region's life science industry. The laboratory addresses basic skills and techniques common to the industry including measuring activity and quantity of proteins, growth and manipulation of bacteria, genetic engineering, polymerase chain reaction and antibody methods. In addition to hands on skills, the course provides context for how and why these techniques are used in the industry.

Workplace relevant soft skills include communication, teamwork, and workplace expectation exercises. This boot camp has been institutionalized by San Diego Miramar College as a course called *Introduction to the Biotechnology Laboratory*. It is taught in a flexible format of 5 days in class with the sixth day on–site in internship. Students are required to do at least 7 weeks of paid internship. The program has produced uniformly positive testimonials:

Intern Mentor: *According to Dr. Terri Quenzer, principal scientist of Pfizer Global R&D La Jolla Laboratories and mentor of multiple LSSI interns, "LSSI's 'Boot Camp' training gives interns a smooth transition from school to industry environment, the interns come in much better prepared."*

High School Student: *Ironically, Summer Puente had loathed her high school Biology course. "The 'Boot Camp' opened my eyes to 'real biology' and how to work in a lab," Summer states. "And the training on 'soft skills' really prepared me for my internship." After completing the training in the "Boot Camp", Summer interned at the Salk Institute in La Jolla. "I came into the program knowing next to nothing about biology. I struggled initially, but by the end I was grabbing all the materials out of the freezers, checking and re-checking them, and basically running the experiment." Summer now plans a life science career.*

College Student: *After completing his internship program with Invitrogen, Billy Chiu decided that he would like to start his career with Invitrogen. "My experience with Invitrogen made it easy for me to get my job, because I already worked there...LSSI taught me that I wanted to work in this industry, and that I wanted to work for a large biotech company."*

Industry Host: *"To bring innovations, creativity, and enthusiastic young people to the company!" Ms. Theresa Schommer responded immediately, while being asked why Invitrogen, employer of hundreds of engineers and scientists with advanced degrees, runs a student internship program.*

TEACHERS: LABORATORY CURRICULUM TRAINING AND INDUSTRY EXTERNSHIP EXPERIENCES

The Amgen-Bruce Wallace Laboratory Program curriculum is the foundation of the teacher and student experience. Consisting of 8 laboratory exercises which can be conducted in typical high school class periods this program exposes the participants to three Nobel Prize techniques and are built around four central paradigms: 1) recombinant DNA technology, 2) gene expression, 3) protein purification, and 4) the polymerase chain reaction (PCR). During week one of the teacher externship training it is expected that teachers gain only a 'student' perspective of these laboratories; curriculum troubleshooting and ongoing support are necessary for teachers to actually implement the labs in their own classrooms.

The industry externships are a key component of the program. The externship experiences give the teachers a context for the science, and exposure to industry professionals who describe both their science and career paths. The diverse industry cluster in San Diego requires that teachers be exposed to at least one large manufacturing site, a large research and development site, a small start-up environment as well as a research institution. The LSSI has been fortunate to have all of these experiences available to the teachers. The hosts provide staff and careful planning that make this an excellent and lasting experience for the teachers. Teacher deliverables from this experience include background research on the hosts as well as a final presentation providing a regional industry overview.

> For high school teacher Malinda Dixon, perhaps the most lasting result of her participation in the LSSI will be the Biotechnology class that she plans to introduce next year—one full year covering biotechnology and its real world uses! Malinda arranged training on the curriculum for all of her colleagues. Her one wish is to have more high school life sciences teachers get involved in the program. "It's absolutely important for high school teachers to participate in the LSSI."

The LSSI Teacher Externship Program is a paid professional

development experience, participants are paid in the form of training stipends. In addition, participants have the option to obtain academic semester units from a California State accredited university.

ONGOING SUPPORT AND IMPLEMENTATION

In 2004 it was recognized by the initial 9 teachers that most professional development programs lack on-going support. In the case of the curriculum provided to teachers in the LSSI, the equipment start up costs are over $40,000 and ongoing supply requirements exceed the entire supply budget of most high school science departments. The teachers partnered with the SCBC and sought funding for 'loaner' equipment and supplies. This support was generously provided by the Amgen Foundation.

However, implementation of the Amgen–Bruce Wallace Laboratory curriculum in the high schools also required an element of labor intensive preparation that most high school teachers lack time for. To address this need, the SCBC employs an Outreach Coordinator to assist teachers with their implementation. In a strategy of learned independence it is anticipated that in three years time, teachers will have developed infrastructure and expertise to leave little need for support other that supplies and occasional loaner equipment.

MEETING THE LOCAL ECONOMIC AND WORKFORCE NEEDS

San Diego, the State of California and the nation have been experiencing a significant and growing crisis in the area of math and science education. As San Diego continues to grow as a world-class hub for biotechnology, science and innovation, the young people growing up in San Diego are not being inspired and prepared to fill the high-skill, high-wage jobs being created within the local economy. The LSSI was created to address this issue head on and at multiple academic levels.

The U.S. Department of Labor's goals were at the heart of the development of the LSSI, as each program was established based on collaborative needs assessments of our region's 21st century workforce. The LSSI focuses on the training and development of upper-level high school, community college and university students as well as high school and community college teachers to create a pipeline of qualified and informed workers for the future. The LSSI programs successfully bridged the gap between industry and academics by providing students with the tools, knowledge and real-world experiences needed to pursue careers and make informed academic decisions. The programs also provided classroom teachers with a more intimate knowledge of the life sciences industry and prepared them to create curriculum and classroom activities that are more relevant, exciting and beneficial to the students they teach.

MEASURES OF SUCCESS

The program model was built to sustain the local workforce. However, there is also an overarching goal to improve scientific literacy by exposing students to the modern biological advances demonstrated by the laboratory activities in the Amgen-Bruce Wallace Biotechnology Lab Curriculum; engage them in the process of scientific discovery, increase scientific literacy, and allow them to develop critical thinking skills.

Effectiveness in meeting this goal is measured in 'hard terms' by how many students and teachers are benefiting from the program. To date, this program has been a great success:

	Summer 2005	Summer 2006	Summer 2007
Student Applicants	38	127	198
Student Internships	13	44	61
Teacher Participants	9	18	24
Companies hosting internships or externships	22	30	25
Approximate Direct Cost	$226,089	$311,917	$260,061
Leverage Match*	$266,836	$396,755	$452,793

* Includes Southern California Biotechnology Center Outreach Coordinator and Funding from the Amgen Foundation to support the Amgen-Bruce Wallace Curriculum implementation

In the first half of 2007, 8 schools implemented the curriculum learned during the LSSI program having an impact on 1,228 students. Three teachers received independent funding for equipment, and over 3,000 students benefited from exposure to enhanced hands-on laboratory curriculum instruction. Several teachers implemented the labs on their own and the effect of the industry externship experience has allowed teachers be more informed when providing career advice to their students. It is estimated that each teacher reaches an average of 189 students per year. As this program expands the number of students reached will grow exponentially. Approximately 6,804 students benefited from the program overall to date (September 2005–June 2007).

Following summer 2007, the LSSI program placed a total of 118 students into hands-on internship experiences at local industry companies and research institutes. Twenty percent of the interns placed in these life science internships continued to work either part-time or full-time for the companies in which they interned. Lastly, evaluation results from both the student and teacher LSSI programs indicate that 100 percent of the participants would recommend the program to their peers.

SUMMARY AND FUTURE CHALLENGES

It is evident that it takes time to develop and implement a program of sufficient scope and depth to attract participants. The relatively slow implementation allowed the stakeholders to develop and refine the program such that a sustainable model can be articulated to other regions. Program metrics show that the program is gaining recognition and acceptance; students who participated in the 2007 program were directly referred by past teacher participants. In addition, past success has prompted an increased interest in the program and overall acceptance.

The LSSI is a unique, replicable program that ensures a local workforce for a major industry cluster in Southern California. The student program sparks interest in life science careers. The teacher training, and the support it provides, allow the education system to better provide industry awareness and improved basic science literacy. Although some key elements of the program have developed 'institutional' support, a major future challenge is the ongoing funding needed to provide training stipends for participants as well as sustain staff members who coordinate and support program activities.

As acceptance of the program continues to grow, we will be careful to not stray from our original goal of better connecting local students and teachers to the life sciences community. We have worked to develop true partnerships where each organization is making an investment in the program and in turn can benefit from its success. We will continue to approach growth and sustainability in a strategic fashion that does not water down the program quality that has become so important to our business and education partners. With each step we take, we will aim to inform, educate and inspire the students we reach, while continuously supporting the teachers who play such a critical role in their development.

We have made a long-term, collaborative investment in this program: an investment that will no doubt pay off for our youth, our education system and the 21st Century workforce that will continue to make San Diego a hub of scientific discovery and region of unprecedented economic growth and opportunity.

ACKNOWLEDGEMENTS

Key Partners
San Diego Workforce Partnership Inc., BIOCOM, Biogen Idec, Invitrogen Corporation, Southern California Biotechnology Center (SCBC) at Miramar College, and The Amgen Foundation

Participating Companies (Hosts for Students Internship and Teacher Externships)
Accumetrics, Alexion Antibody Technologies, Anadys Pharmaceuticals, Arena Pharmaceuticals, Assure Controls Inc., Biogen Idec, BioServ Corporation, Burnham Institute for Medical Research, Conatus Pharmaceuticals, Conservation and Research for Endangered Species (CRES), The Dow Chemical Company, eStudySite, Genentech, Genomatica, Genoptix,

Gen-Probe, Invitrogen Corporation, Isis Pharmaceuticals, Karl Strauss Brewing Company, Nanogen, Pfizer, Salk Institute for Biological Studies, San Diego State University Labs, Santarus Inc., SCBC@Miramar College, SGX Pharmaceuticals, Sharp Chula Vista Medical Center, Skin Medica, Sunrise Science Products, The Scripps Research Institute.

Florida Community College at Jacksonville's Biotechnology Laboratory Technology and Bioinformatics Programs

R. Kevin Pegg

R. Kevin Pegg is Professor of Biotechnology at Florida Community College at Jacksonville. Kevin can be contacted at *rpegg@fccj.edu*.

Florida Community College at Jacksonville offers Associate of Science degrees in biotechnology, and certificates in bioinformatics. Both curricula are intensive, hands-on programs designed to prepare graduates for the expanding job market.

Biotechnology is a wide-ranging field encompassing individual researchers analyzing single-gene mutations, as well as engineers operating house-sized purification systems. Graduates of our A.S. program prepare for employment in academic laboratories as molecular biology research assistants, and for manufacturing environments as technicians using molecular techniques on an industrial scale.

The curriculum is designed to assure potential employers our graduates have short learning curves in either environment. To achieve this, our program emphasizes universal and scalable laboratory skills. Solution formulation is one example. Students master preparing buffers, reagents, media, admixtures, colloids and gels at scales from multiliter down to milliliter. Further, in our core curricula, students formulate all the solutions needed to isolate and analyze DNA, RNA, and proteins. For some Standard Operating Procedures (SOPs) they prepare a dozen solutions before performing the actual experiment. In one project, our students "clone themselves" by isolating a small amount of their individual DNA. They then insert their genes into bacteria making a simple genetic library. However, before the constructing the library the class produces four kinds of bacterial growth media and nine different buffers and reagents for their molecular work.

This training contrasts with typical college laboratory experiments where students use prepared media or kits, and perform only the last steps of a single procedure. In the Biotechnology Laboratory Technician degree program trainees carry out all parts of a method, not just the endpoint of one recipe. Once in the workplace these skills and insights ensure success in any environment.

The biotechnology A.S. degree program comprises 60 credit hours of both general education courses and laboratory-based classes emphasizing technician skills. The bioinformatics certificate provides information technology professionals with the skills needed to understand the unique requirements of genetics and molecular biology industries.

PROGRAM OF STUDY

To earn the Biotechnology Laboratory Technology Associates Degree students complete 16 credit hours of general education courses—english, math, humanities, and computer literacy—plus 23 credit hours of chemistry, statistics, biology and microbiology. There are 22 credit hours of biotechnician classes, including a required internship working in a biotechnology laboratory. The program is designed around a two-year pathway of full-time course work.

BIOTECHNOLOGY COURSES

Introduction to Biotechnology
A career in biotechnology is the main theme of this course. Following a brief introduction to molecular biology, the course focuses on career choices such as pharmaceutical, pharmacognosy, diagnostics, forensics, agricultural, and aquaculture biotechnology. This is a one credit-hour course meeting four times that includes hybrid discussions online. Students have the option of a graded

Table 1: Program of Study

Recommended Biotechnology Laboratory Technician (2199) Sequence Of Courses		
Biotechnology Laboratory Technician Year 1		
Fall	Spring	Summer
Intro to Biotechnology BSC 1421 1 credit	Biotechnology Methods I BSC 2420C 4	Biotechnology Methods II BSC 2427C 4
Intro to Biotechnology Methods BSC 1404 3	Intro to General Chemistry CHM 1025C 4	Gen Chemistry & Qual I CHM 2045C 3
English Composition ENC 1101 3	Humanities 3	Microbiology MCB 2010C 4
College Algebra MAC 1105 3	Social and Behavioral Sciences 3	Elementary Statistics STA 2023 3
Principles of Biology I BSC 2010C 4		

Biotechnology Laboratory Technician Year 2		
Fall	Spring	
Protein Biotechnology / Cell Culture BSC 2419C 4	Internship BSC 1943 3	
Intro to Bioinformatics BSC 2435 3	Gen Chemistry & Qual II CHM 2046C 4	
Human Biology BSC 2020C 4		

paper or presentation on a career interest of their choosing. The course is typically taught in the biotechnology laboratory to give students exposure to the instruments and apparatus they will train on. A fully online version of the course is under development to increase exposure to biotechnology as a career.

Introduction to Biotechnology Methods
This is the gateway course to the other laboratory classes in the biotechnology series. Laboratory orientation, volumetric measurements, gravimetric measurements, and molecular biology, along with skill building exercises, are the main themes of the early portion of this class. All our courses have macromolecule themes. *Introduction to Biotechnology Methods* has a primary focus on DNA; subsequent classes focus on RNA or proteins. In this course students construct a simple genetic library of their DNA. In the pro-

cess they learn to make all solutions for gels, reagents and media, to isolate DNA from bacterial plasmids and their own human epithelial cells, and to work with restriction enzymes. Students gain experience in laboratory technique and methodology, familiarization with biological cloning, and learn precision and accuracy with weight and volume measurements. Students give two PowerPoint presentations on metrics, DNA isolation, and cloning topics.

Biotechnology Methods I
There is an emphasis on observation, and data collection and manipulation in this course. The main theme is isolating and working with RNA. Students isolate total RNA from three species of *Bacillus*. Students grow and maintain all bacterial strains, as well as make all media and buffers used in the experiment. Spectroscopy techniques are used to calculate total RNA from each species, and spectroscopy and gel analysis are used to determine which samples are cleanest and have the highest nucleic acid concentration. There is a formal lab meeting to choose the best isolates for subsequent conversion to cDNA using reverse transcriptase, followed by *in vitro* amplification with PCR. Amplification fragment length polymorphism (AFLP) maps are then generated for the specimens. Southern blots and oligonucleotide arrays are taught in this section as well.

Feedback from our internship preceptors suggested students needed more training in spreadsheets, and this set of experiments was modified to make extensive use of Microsoft Excel. On completion of this section, students have isolated and purified both DNA and RNA, amplified DNA by both *in vivo* and *in vitro* methods, run a variety of gels, made dozens of solutions, performed analytical spectroscopy, and participated in a group meeting that resulted in "go/no go" decisions based on data in spreadsheets. Students have two presentations on topics in macromolecules and molecular biology.

Biotechnology Methods II
This course emphasizes immunochemistry and protein purification. Students isolate peroxidase enzyme from horseradish root

using salting-out techniques, and perform enzyme assays and poly-acrylamide gel electrophoresis to determine purity. Additionally, students complete sections in immunochemistry by performing microtiter plate enzyme linked immunosorbent assays (ELISA), as well as developing a pregnancy test-style solid-phase immunoassay test strip using colloidal gold antibodies. They learn colloidal gold nanoparticle coating techniques and immobilization of antibody within a porous-phase support. Western blots are covered in this section.

Aligning and using a phase contrast microscope along with epifluoresence techniques are also part of this course. Lectures relating light and spectroscopy are given and students "build" a spectrophotometer using an Ocean Optics fiber optic spectrometer, ultraviolet lamp, and cuvette holder.

Lectures focusing on digitization of data are coupled with practical exercises. Students use an analog-to-digital (A/D) converter to change analog voltage electric signals from light and temperature sensors into digital signals, and then store the digital data in spreadsheets for graphing and analysis. Students give two presentations during the course on topics in immunochemistry, microscopy, and digital data.

Tissue Culture and HPLC

This is the most intensive technical course we teach. Students are expected to demonstrate advanced skills and genuine insight into techniques used in the lab before taking this course. Students train on simulated cell passage techniques inside and outside the biological safety cabinet. We use plant tissue culture with explants from African violet leaves to hone sterile technique skills. Students must pass a sterile technique lab practical before they receive chinese hamster ovary (CHO) cells. They are required to subculture the CHO cells in antibiotic-free media continuously for 6 passages without contamination. Additionally, students generate several volumes of cells for cryogenic freezing and thawing, and for cytogenetic studies.

Students produce karyotypes showing their techniques did not alter the gross nuclear morphology of cells during the period the cell lines are in their care. Students also document their work using photomicrography with images from phase contrast microscopy. Fluorescent imaging, Sister Chromatid Exchange (SCE) using chromosomal labeling, Giemsa dye banding, and Fluorescent *In Situ* Hybridization (FISH) techniques are also covered. Students perform a cell synchronization experiment arresting cells during interphase, then recovering the cells for a coordinated mitosis.

We also use this phase of the course to emphasize laboratory etiquette. Since most laboratories share equipment and resources, students work with the campus laboratory managers to schedule access to the hoods and prep areas outside regular class times. In this case, they work directly with the lab manger, not the instructor, and are given sample memos for requesting resources. Labs are complex social environments, and this portion of the training is designed to provide students a framework for interactions in a real lab environment, rather than a teaching lab.

High Performance Liquid Chromatography training is conducted during this phase. Students learn theory, buffer preparation, sample preparation, and data analysis using reverse phase and gel permeation HPLC. Students have two presentations in topics of cell culture and HPLC.

Bioinformatics

This course is taught in a computer lab; each student has a terminal and the instructors' screen is projected for class view. Students follow along on their terminals entering examples and performing searches.

Most students need substantial remedial training in database searches at the beginning of the course. As they are already familiar with MS Excel from the laboratory coursework, they train in converting spreadsheets into MS Access databases, and learn to generate search queries and reports in Access. We rely on Microsoft's sample Northwind database training tools for this section. Training the use of database Boolean operators is conducted by teaching students how to perform WEST searches (Web-based

Examiner Search Tool) in the patent databases at the United States Patent Trademark Office. Students learn subject matter searches, inventor searches, and restricting record sets.

After training on search techniques the course moves to molecular biology case studies using BLASTp and BLASTn, and databases in the National Center for Biotechnology Information (NCBI). The Online Mendelian Inheritance in Man (OMIM), Probe, and PubMed databases are particularly useful. This is a hybrid course with a significant quantity of homework for practice outside class to ensure students develop strong computer skills. There is a final practical with students at computer terminals interpreting case studies, using online databases to find the answers, and writing short essays on the case.

Ancillary courses

The 2-year plan is designed for students pursuing an A.S. degree as a ticket to work in a laboratory. Portions of the course can be tailored for students pursuing advanced degrees at other institutions. For example, some students may substitute anatomy and physiology courses for the human biology requirement. For some students already possessing advanced degrees in closely related fields a one-year course plan can be mapped out.

COURSE REQUIREMENTS FOR THE BIOINFORMATICS ADVANCED TECHNICAL CERTIFICATE

This program is primarily designed for IT professionals who have completed the Database Technology A.S./A.A.S. degree and have an interest in learning more about genomics and proteomics. Students take fourteen credit hours of *Principles of Biology, Introduction to Biotechnology, Introduction to Biotechnology Methods, Bioinformatics,* and *Introduction to Perl Programming* on the biotechnology side. On the IT side ten credit hours from offerings of Linux/Unix, Linux Network Administration, Oracle Database courses and HTML/XML electives complete the certificate.

CONTINUING COURSE EVOLUTION

Our biotechnology courses are constantly revised through feedback from our advisory board and internship preceptors, interactions with other programs in the state, and from follow-up with our graduates. The advisory board is a mix of science faculty from the college, faculty from other local educational and research institutes–such as the Mayo Clinic at Jacksonville, and local biotechnology and biomedical company executives such as Medtronic, Inc.

The advisory board meets twice annually to review changes in the coursework structure, discuss the progress of our students, and to inform our program heads of changes in the local research and industrial climate that might affect our program.

Especially valuable are comments from laboratories where our students perform internships. Many small suggestions—such as increasing emphasis on notebook record keeping and use of spreadsheets—aid our students transition from academic learners to laboratory workers.

The FCCJ program has ties to similar programs around the state. Most notably are Santa Fe Community College (SFCC) in Alachua, Florida and Indian River Community College (IRCC) in Ft. Pierce, Florida. These interactions are both informal through faculty discussions, and formal through tri-annual meetings via the Florida Workforce biotechnology consortium called BioTEC. While the individual college programs differ in the types of training and local focus there is considerable value in hearing what works and what doesn't from other schools.

At least once a year we bring together working graduates and interning students, in a single forum with current students, where they can exchange information and talk freely with one another. Perhaps the greatest feedback of all comes from following the achievements of our graduated students.

INTEGRATION WITH STATE GOALS

FCCJ's curricula are tightly integrated with state goals for ensuring growth in biotechnology in Florida. Florida's growth and commitment to biotechnology as a business can be traced to the early 1980s at the inception of the biotechnology industry. The state now has two concentrations of pharmaceutical-based biotechnology businesses. One cluster of industries is located in the center of the state around the University of Florida, associated with the biotechnology incubator in Alachua. A second concentration of companies and institutes occurs as a biotechnology corridor running diagonally across the southern center of the peninsula, from the Miami-Dade area through Orlando to Tampa. The biotechnology corridor is a state-sponsored initiative, with the state of Florida investing nearly $1 billion to bring Scripps, Torrey Pines Institute for Molecular Studies, and The Burnham Institute for Medical Research to the south Florida region.

FCCJ's biotechnology program participates in the statewide Florida Workforce biotechnology training initiative. Operating through the BioTEC cluster center the program currently consists of the three community colleges with active biotechnology training programs—FCCJ, SFCC, and IRCC—with the University of Florida. Regional meetings are held around the state to encourage academic and industry cooperation in biotechnology workforce training, and to foster development of new programs at other regional campuses.

In 2007 the consortium of community colleges launched the first statewide course entitled: *Biotechnology: Working in a Regulated Environment*. Drawing nearly two hundred applicants statewide, the program issued Florida's first 40-hour Certificates in Biotechnology using a hybrid format of on-line and hands-on laboratory instruction. This initiative will, one day, yield Florida-wide accreditation of biotechnician skills courses in sync with industry's needs. In the summer of 2007 twenty-seven biotechnicians received this novel certificate at FCCJ.

INTEGRATION WITH THE LOCAL COMMUNITY

Northern Florida, where FCCJ's campuses are located, has biotechnology businesses ranging from small entrepreneurial efforts in genetic diagnostics to mid-sized biopharmaceutical companies. The area also has medical device manufacturers and clinical research enterprises such as the Mayo Clinic, Jacksonville. North Florida is home to several large-scale businesses with biochemical and molecular components, such as marine fisheries import/export operations centered at the port of Jacksonville.

At this writing the FCCJ Biotechnology program is working on a joint academic/industry partnership with a consortium of local seafood processors to develop a nonprofit Institute for Food Safety. Led by Beaver Street Fisheries, Incorporated, on the industry side, the novel institute will engage in compliance-testing for regional seafood processors. Beaver Street Fisheries is a local success story, a family-owned business that began as a small fish shop and grew into a major importer of seafood from over 50 countries. Limiting future growth of this regional industry is a need for local third-party analytical testing of products to meet federal quarantine-release guidelines.

The development of the institute will aid growth of an industry complimentary to north Florida's role as a major seaport. The institute itself will provide traineeships, internships, and jobs for our biotechnology graduates. Jobs affiliated with port operations in Duval and surrounding counties can follow expansion of the seafood import/export businesses past the current regulatory bottleneck. As envisioned, the institute will operate as a 20,000 square foot state-of-the-art laboratory and classroom co-located with cold storage facilities. When not used for biotechnology training, the classrooms can provide local workers with opportunities to learn English as a second language and improve their reading skills. This operation will serve as a model for similar industry/college partnerships in the future.

FCCJ's biotechnology A.S. degree is vertically integrated with area high school programs as a source of new students; and with state and regional four-year colleges and universities to provide graduates with additional educational outcomes. The biotechnology program is part of the colleges' career pathway program whereby local high school students earn college credits in their junior and senior years. The college works with local magnet high schools with a science/biotechnology focus to encourage students into fields in science, technology, engineering, and math. Our course credit structure is designed to be transferable to state universities. The college uses articulation agreements with regional universities to provide students with post-associates transfer routes for those students continuing their academic interests.

The biotechnology program demonstrates a diverse appeal, attracting students in their teens using the program as a stepping stone to a higher degree, as well as retraining seasoned teachers and technologists looking for a second career. Graduates have moved on to bachelor's programs in molecular biology, found work as research technicians, or moved into biomanufacturing or regulatory testing areas. The program is growing with the local area and the state as biotechnology moves into the mainstream as an employer.

ADVANCED TECHNOLOGY CENTER

FCCJ has five campuses and six educational centers located throughout the north Florida region. Including international online programs, the college has an enrollment of 75,000 students, and the college campuses host a variety of meetings, conferences and cultural events impacting every facet of our local community.

The biotechnology laboratory and classrooms are located in the Advanced Technology Center (ATC) on the college's downtown campus. The ATC is a $25 million, 121,000 square foot leading-edge facility offering programs of study including production processes; industrial electricity and PLC; information technology, including networking and IP telephony; drafting CAD; transportation technology; and, biotechnology and biomanufacturing. All programs at the ATC are industry focused and often led by participating local businesses.

The biotechnology program is housed in a specially-built 3,000 square foot wet lab "smart" classroom with stations for 24 students. The facility can be configured for individual workstations, or for group projects. Analytical equipment is available in the classroom. An embedded prep room houses formulation equipment, cryogenic freezers and an autoclave. Wireless Internet through the college servers is provided in all areas. Bioinformatics is taught in a separate computer laboratory in the ATC.

PROGRAM HISTORY

The biotechnology laboratories and curricula were part of the overall development of the ATC. FCCJ's biotechnology programs were started by the college president, Steven Wallace, in 2000, at the same time funding was sought for building the ATC. A funding framework for the ATC and affiliated programs was constructed through federal and state grants, and college programs. Primary responsibility for carrying out the biotechnology initiatives fell to campus president Edythe Abdullah and college executive vice president Donald Green. Kathryn Birmingham, dean of liberal arts and sciences, was responsible for the nuts-and-bolts aspect of the program's genesis. Although many faculty and staff members aided development of the program, Professor Inga Pinnix was the lead faculty member during the startup and deserves credit for shepherding this project through the numerous committees and credentialing bodies within the college.

Additional funding and growth came from several college initiatives during the start-up and growth phases. Two initiatives that currently affect our program are the statewide Workforce Education Cluster Center for Biotechnology and the college-wide Florida Life Sciences Think Tank.

In 2003 a $1.2 million competitive grant through the State Workforce Board for biotechnology education established a consortium of academic and industry partners to develop statewide web-based curricula. The founding partners were Embry-Riddle

Aeronautical University in Daytona Beach, Florida Community College at Jacksonville, Indian River Community College in Fort Pierce, the International Society for Pharmaceutical Engineering, Santa Fe Community College and the University of Florida in Gainesville. The Biotechnology Training and Education Council, or BioTEC, has expanded and is now headquartered at UF through a subsequent half-million dollar grant in 2006 and a $200,000 grant in 2007. The Cluster Center has established the Employ Florida Banner Center in partnership with BioFlorida, the state's industry association, as a driving force in biotechnician workforce education.

College-wide, advancement of the biotechnology program is part of the Florida Life Sciences Think Tank project. Initiated with the drafting of a college-level grant proposal by Professor Lourdes Norman, the Think Tank seeks to anticipate future needs of all the life sciences programs, and assure the facilities and curricula meet those needs. Numerous faculty and staff have participated in advancing the project, primary responsibility for implementation is under college Vice President Nancy Yurko. Several implementation strategy meetings have been held on- and off-campus. Notably, the 2005 meeting in conjunction with the Virginia community colleges consortium featured guest speakers on the future of biotechnology, and produced a comprehensive report and recommendations.

STUDENTS

In our program, diversity makes a "typical" student difficult to define. The age spectrum is from teens seeking a first degree to those in their fifties wanting a second career. The classes are about two-thirds female. Nearly a third of the students have migrated to the U.S., and English is a second language for them. All ethnic groups are represented, without a trend.

Most students enrolled in other programs before choosing biotechnology as a career path. Typical prior training includes nursing, general biology, and medical technology; however, we also have students with backgrounds in aviation, videography, and fine arts. About two-thirds are seeking employment immediately on graduation, the rest plan advanced degrees.

Students who do well in the program, and achieve early success on graduating, had a keen interest in molecular genetics before entering the program. Nearly all were self-studying for years before starting formal training. There is a strong link between success in the program and long-term interest in health sciences.

STUDY IN CONTRAST

Two students in the inaugural cohort illustrate the diversity and range of students in our program. One is a female with a school-age child; she is an immigrant with English as a second language. She has a biology master's degree from her native land, which translates to a bachelor's degree in the U.S. She had no prior laboratory work experience, either in her country or here, and the curricula in her home country emphasized textbooks with a few demonstrations, and no actual laboratory skill-building exercises.

The other student is male; the quintessential all-American boy. He started at the college after graduating from a local high school and entered the biotechnology program with just a few of the general education requirements behind him.

Despite the contrast in their backgrounds, both graduated at the top of the program, both obtained near perfect scores on a comprehensive bioinformatics practical, and both worked very hard perfecting laboratory skills. Both completed excellent internships in world-class laboratories, and both are successful post-graduation.

Based on our experience, the only predictors for student success in this field are attention to detail and innate desire to work in a laboratory environment. Accordingly, we look for students who are willing to work hard developing the manual skills needed to manipulate microfluids and perform basic laboratory tasks with precision and accuracy. A desire to understand the nature of macromolecules is essential.

Recruitment

Our pool of enrollees comes in part from advertising the program via mixed media, somewhat from college degree promotion fairs, and mostly from students actively looking for a path to working in a laboratory. Despite print, radio, and television ads about the biotechnology program, most students say they found it by randomly scanning the school catalog for programs meeting their interests. We receive only a small percentage of new contacts per year from our in-school career counselors.

A large percentage of our students are cross-training to increase their employment options. Several students each term are dual-enrolled in the Biotechnology A.S. program and B.S. programs in biology at local four-year institutions. Each of these students found the biotechnology program because they were specifically looking for skills-training not provided by traditional academic science programs.

Well over half of the students who contact the program are not familiar with biotechnology. Two large groups include prospective students confusing biotechnology with naturopathy (use of plants as herbal remedies), and those looking to find a job in crime-scene forensics because of infatuation with televised drama shows.

While we do not have a formal application process, a considerable amount of time is spent discussing biotechnology career options with prospective students. Although there is a web site for the program, students usually call for more information. Individual interviews with lab tours are the most successful way of obtaining serious students.

Campus and college-wide career fairs have not been useful so far as a recruitment tool. "Open houses" where students and parents can visit the laboratory for a tour attract considerable interest; however, open houses do not enhance enrollments. We engage in high school outreach, hosting science classes and groups of seniors, and meeting with high school guidance counselors as well. We expect these events will increase awareness of biotechnology and perhaps, in the future, translate to increased enrollments.

Lack of local knowledge about biotechnology is a barrier to recruitment. Few residents are even aware biotechnology exists in this region. It is taking a very concerted effort to reach wider audiences. Local companies and institutions that would benefit from expanded training are just now becoming sufficiently well-organized to sponsor some promotion. The statewide biotechnology industry advocating body is BioFlorida, and a chapter in northeast Florida now organizes several local events each year. The program at FCCJ will likely grow from current enrollment levels in synchronization with growth in the local industry. College collaboration with the Jacksonville Chamber of Commerce and county government is underway with the view of attracting a large biotechnology concern to this area that can help expand employment in this field.

The biotechnology program has benefited from local media. One class was featured on a local National Public Radio program, *In Context*, while performing a cloning experiment and another class assisted a local TV station with a story on microbial contamination of cell phones. Another magazine story on the emerging biotechnology job market featured interviews with the students. The students enjoyed the celebrity, and the coverage did increase awareness of the program.

SUMMARY

Florida Community College at Jacksonville's Biotechnology Laboratory Technology Associates Degree program is a hands-on workforce program designed to meet local, statewide, and national needs for laboratory assistants. Graduates are thoroughly grounded in basic laboratory skills and trained in advanced molecular biology techniques. Students are acclimated to both research and industrial environments.

The biotechnology program is vertically integrated with local high schools on one end for recruitment and outreach, and with four-year colleges and universities on the other end to provide additional educational avenues for our graduates.

The program participates in state initiatives, including the Florida Workforce BioTEC consortium. The program is tied tightly to the needs of local biotechnology employers in industry, research and government.

Biotechnology is the future; already producing new medicines, biotechnology will expand geometrically over the next century to impact everything from food to fuels. At FCCJ we are endeavoring to ensure our program, like biotechnology itself, meets the needs of the 21st century.

UC Davis Biotechnology Program
Judith A. Kjelstrom and Denneal Jamison-McClung

Judith A. Kjelstrom, Ph.D., is Director of the University of California, Davis, Biotechnology Program. Judith can be contacted at *jakjelstrom@ucdavis.edu*.

Denneal Jamison-McClung, Ph.D., is Assistant Director of the University of California, Davis, Biotechnology Program. Denneal can be contacted at *dsjamison@ucdavis.edu*.

The University of California, Davis, has been an innovator in biotechnology research, education and training for over 20 years. In 1986, the UC Davis Biotechnology Program[1] was founded to assist in the organization of university activities related to biotechnology and to coordinate with related efforts on the Davis campus.

The Program's missions include:
1. Promoting and coordinating the development of biotechnology and biotech–related research on the campus
2. Assisting with the development of new and improved facilities for biotechnology research
3. Promoting research interactions between faculty and private industry, and public agencies
4. Recommending and implementing curriculum development and training in biotechnology
5. Serving as an informational and educational resource on biotechnology for the campus and the public

Currently, UC Davis is the only UC campus with a stand-alone biotechnology program. The program serves a critical role in

---

1   *www.biotech.ucdavis.edu*

bridging academia with industry and government, fostering critical partnerships to enhance the education and training of students and researchers, and supporting early-stage biotech companies in the region. Technology brokering is critical to moving ideas from the bench to the market.

New projects within the Biotechnology Program are developed to address critical needs in biotechnology research and education/training. Communication with industry is an important element in defining entry level and advanced skill sets required in training the biotech workforce. The Biotechnology Program has a can-do spirit and is very successful in developing pilot educational programs involving multiple partners from academia, industry and government. To support and expand pilot programs with significant positive outcomes, dedicated partners in both academia and industry work with program personnel to identify and apply for sources of institutional support (i.e. state and federal grants, industry donations, private donations, etc.).

Base funding for the Biotechnology Program comes from the Office of Research and office space has been offered by the dean of the College of Biological Sciences. Operational funds are generated through grants, sponsorships and registration fees. Support for the Biotechnology Program is sustained by demonstrating value to the campus and the region. The program builds value by partnerships with UC Davis faculty and regional colleges to secure grants and equipment. Outreach efforts in the community are a powerful vehicle to showcase the cutting edge research and education at UC Davis. Summer short courses and *Train the Trainer* workshops fulfill vital technical training needs for the campus and the region.

EDUCATION & TRAINING

The Biotechnology Program is involved in a number of educational/training programs and serves as the administrative home for many of them:

1. Designated Emphasis in Biotechnology (DEB)

graduate program for PhD students[2]

2. NIH–NIGMS Training Program in Biomolecular Technology for PhD students

3. NSF IGERT-CREATE (Collaborative Research and Education in Agricultural Technologies and Engineering) Training Program for PhD students

4. Howard Hughes Medical Institute's Med into Grad Training Program entitled Integrating Medicine into Basic Science (IMBS) for PhD students

5. Advanced Degree Program (ADP) for corporate employees: a PhD program for the working professional in the biotech industry

6. Summer Short Courses: week-long, cutting-edge lecture/lab courses on Flow Cytometry, Proteomics, DNA Microarray Techniques, and Advanced PCR, for faculty, graduate students and industry scientists and "Train the Trainer" workshops in Biotech & Bioinformatics for teachers

7. Entrepreneurship Training

8. BioTech SYSTEM: a K-14 educational consortium in the region to support education, training and mentoring. The Teen Biotech Challenge is one of the major activities of the consortium.

THE DEB GRADUATE PROGRAM

UC Davis was recently awarded a competitive five-year renewal of the prestigious NIH Training Grant in Biomolecular Technology (T32 GM008799; 2007-2012), in recognition of the quality of multidisciplinary research and training provided by the campus. The DEB is the formal training program for the grant and all fellowship recipients must be part of the DEB. The grant is under the directorship of Prof. Bruce Hammock, Department of Entomology and the Cancer Research Center, with co-directors Professor Karen McDonald, Department of Chemical Engineering and Associate Dean of the College of Engineering, and Dr. Martina

2   *www.deb.ucdavis.edu*

Newell-McGloughlin, UC Systemwide Biotechnology Program and Department of Plant Pathology. Dr. Judith Kjelstrom is the program manager for the NIH Training Grant in Biomolecular Technology and the DEB graduate program.

The name, biomolecular technology, was chosen to reflect the emphasis of the training program on areas of scientific endeavor characterized by the following three elements:

1. Emphasis on the analysis of model systems of obvious significance to medicine and biotechnology
2. Synthesis of information and research approaches from disciplines such as cellular physiology, genetics, physical biochemistry, and chemical engineering
3. Translation of biological information into a quantitative framework

Through these three foci, the program provides well-coordinated, multidisciplinary training of pre-doctoral graduate students in critical areas of biotechnology research that address public health. It also provides an administrative structure for creating interdisciplinary research environments that integrate basic biological science and engineering disciplines, as well as academic and industrial experiences. The program is designed to recruit and support pre-doctoral trainees who show exceptional promise, coupled with the drive to reach out across disciplines and forge new research directions in biotechnology.

As stated above, the formal training program for the NIH Training Grant in Biomolecular Technology is the Designated Emphasis in Biotechnology (DEB) graduate program. The DEB is an inter-graduate group program, housed within the offices of the UC Davis Biotechnology Program. The DEB allows PhD students to receive and be credited for training in the area of biotechnology. A student may be a member of the DEB program even if he/she is not funded on the NIH Training Grant in Biomolecular Technology.

The DEB program supplements a student's PhD curriculum and those completing the DEB program will obtain an official designation on their diploma and transcript indicating a qualification in biotechnology. For example: Doctoral Degree in Microbiology with a Designated Emphasis in Biotechnology.

The DEB Mission is:
1. To provide well-coordinated, cross-disciplinary training of graduate students in critical areas of biomolecular technology research
2. To promote interdisciplinary research environments that integrate basic biological science, engineering and computational disciplines
3. To allow cross-disciplinary training and trainee experience in a biotechnology company or a government laboratory

Pre-doctoral students come from a wide array of disciplines: applied science; biochemistry & molecular biology; biological & agricultural engineering; cell & developmental biology; biostatistics; chemical engineering; comparative pathology; entomology; genetics; microbiology, plant biology; plant pathology; statistics; etc.

DEB CURRICULUM

The DEB keystone course is *Biotechnology Fundamentals and Applications (MCB 263)*, which is team-taught by molecular biology and engineering faculty. During the course, students must work in interdisciplinary teams, composed of a biologist and an engineer, to complete a class project. Working in interdisciplinary teams is very challenging! Bioethics is a critical component of biotech education, so *Scientific Professionalism & Integrity (GGG296)* or an equivalent course is required.

Three quarters of attendance is required at the *Current Progress in Biotechnology (MCB 294/ECH 294)* seminar class, offered in fall, winter and spring quarters. This eclectic seminar series covers a wide range of biotech-related topics, as well as the business of biotech. Lunch is hosted after the seminar, which enables faculty and students to meet speakers, which are recruited from both industry and academia. From an educational standpoint, post-seminar lunches offer valuable opportunities for students to hone their professional networking and communication skills.

The seminar course, *From Discovery to Product: An Introduction to Industrial Biotechnology (MIC 292)* is offered every other year and consists of a series of seminars from scientists and engineers from Novozymes, Inc. The course includes a tour of the Novozymes R&D facility and gives DEB students a good overview of the operations of a biotechnology company, including knowledge of industry-related skill sets. Both *MIC292* and *MCB296/ECH294* seminars are open to the public and they offer UCD faculty and students the opportunity to interact with industry scientists to arrange research collaborations, internships, graduate fellowships, etc.

The highlight of the DEB curriculum is the *Research Internship (MCB 282)*. DEB students must complete a minimum of 3 months at a biotechnology company, national laboratory or cross-college laboratory. This professional networking is a win-win opportunity for the campus, as well as the company. Attendance at the annual Biotechnology Training Retreat (many industry partners attend) & monthly informal Pizza *Chalk Talk* Seminars (student and mentor discuss their research) is also expected. In both of these venues, students have the opportunity to sharpen their scientific communication skills through oral and poster presentations.

The DEB graduate program is somewhat unique in that it stresses both academic expertise and "social awareness." To create effective cross-disciplinary teams, students must value the so-called "soft skills." Deep, narrow expertise, gained through doctoral research, must be balanced with broad, global perspectives to be an effective leader in the 21st century. Recommended reading includes the book, "Social Intelligence: The New Science of Human Relationships" by Daniel Goleman (Bantam; 2006).

In addition to offering academic courses, the DEB graduate program provides guidance and assistance in:

1. Locating internships
2. Strategic career explorations
3. Creation of oral presentations
4. Composing cover letters and curriculum vitae, especially for non-academic positions
5. Networking with industry & government scientists, as well as business leaders

A new NSF training grant, linked to the DEB, was funded in the summer of 2007. The Integrative Graduate Education and Research Training Program (IGERT) – CREATE (Collaborative Research and Education in Agricultural Technologies and Engineering) will train PhD scientists and engineers in novel areas of plant-based technologies. Professor Karen McDonald, Chemical Engineering department, will direct the grant. Dr. Kjelstrom will serve as senior personnel and Dr. Jamison-McClung will manage the program. All PhD students in this training program will also be members of the DEB graduate program. The multidisciplinary research partnership of biologists and engineers, envisioned by the directors, will focus on genetic engineering of plant biosynthesis to permit efficient production of products that include enzymes, biofuels, and vaccines. Industry internships will take place in both the United States and Ireland.

The Howard Hughes Medical Institute's Med into Grad Training Program entitled Integrating Medicine into Basic Science (IMBS) is another training program that is linked to the DEB (although not formally). It is led by the UC Davis School of Medicine in partnership with the Biotechnology Program. Executive Associate Dean Ann Bonham is the director and Dr. Kjelstrom serves as a co-director for recruitment and retention. This training program exposes PhD students to issues in clinical medicine, in order to clearly link basic research to the clinic. A number of PhD IMBS scholars are also members of the DEB graduate program. Translational research is a common theme in both programs, as well as a connection to industry. Effective "bench to bedside" research must include industry partners. Visits to Genentech and Lipomics Technologies (focus on metabolomics) are two examples of industry interactions that help IMBS scholars gain perspective on translational approaches to research in human health.

INDUSTRY PARTNERS ARE A KEY ELEMENT OF THE DEB

DEB students may interact with industry scientists at the *Current Progress in Biotechnology* seminars, as well as the annual Biotechnology Training Retreat in the Napa Valley. The Retreat is a day of scientific talks by biotech fellows and industry partners, poster presentations and a gourmet lunch. The DEB students are encouraged to introduce themselves to the industry participants and to distribute their CVs, if they are interested in a possible internship or future employment opportunity. These networking activities have led to internships and job offers.

Over the last 16 years, including the time prior to the establishment of the formal DEB program in 1997, over 100 PhD students have been placed in a variety of biotechnology companies for three-to-six month industrial research internship experiences. Biotech companies participating in the internship program include: AgraQuest; Agilent; Alza (a Johnson & Johnson company); Amgen; Amyris; Bayer; Berlex (now part of Bayer); BioMarin Pharmaceuticals; Celera AgGen; Chiron (now Novartis AG); DuPont; Exelixis; Genencor; Genentech; ICOS; Maxygen; Monsanto, Calgene Campus; Novozymes, Inc.; Novartis AG-Vacaville; Scios (now part of Alza); Roche Biosciences; Ventria Bioscience and a few others.

Most companies have given *Current Progress in Biotechnology* seminars and have presented at the annual Biotechnology Retreat. Of the more active industry partners, some have donated a biotechnology graduate fellowship to supplement the NIH Biomolecular Technology Fellowships.

What do Industry Partners Gain from Internships?
- Access to highly talented creative researchers
- Opportunity to gain inside track on future employees
- Through students, further collaboration with university scientists
- Participation in the Annual Biotech Retreat to meet UCD scientists, DEB students, and other company scientists
- Potential access to UC facilities through the collaboration
- Opportunity to participate in weekly campus seminars

What do Students Gain from Internships?
- Opportunity to work in highly creative non-academic environment and participate in industry seminars
- Opportunity to participate in focused team approach to defined research goals
- Opportunity to use equipment and facilities not available on campus
- Access to potential employment opportunities
- Discover the type of environment that would be appropriate for future career

HOW DO WE MEASURE SUCCESS?

The growth of the DEB program from 20 to over 150 pre-doctoral students in 8 years is impressive. PhD students definitely find value in the program. According to surveys of past and current DEB participants, students do not feel burdened by the required coursework and participation in DEB activities. The DEB graduate program provides many unique courses and opportunities for pre-doctoral students, which are fully outlined on our program website. The website is a great recruitment tool and many DEB students state that they came to UC Davis for their graduate training because of the DEB program.

The academic and professional quality of DEB students is very high. Therefore, biotechnology companies are anxious to have DEB interns. The emphasis on academic expertise, teamwork and social skills are greatly valued by industry. In the last two years, there has been a dramatic increase in the number of companies taking interns. For example, Genentech created a formal MOU to state that they would take at least three DEB interns, as well as provide a Biotech Fellowship every year. Due to the success of this partnership, Genentech plans to increase the number of interns in 2008.

Another measure of DEB program success is the increase in the number of training grants to provide fellowship support. The success of the DEB was a key factor in obtaining renewed NIH funding for the Training Grant in Biomolecular Technology. In addition, institutional support from the campus Deans and the Office of Research, as well as fellowships provided by biotech companies are evidence of value. The new NSF IGERT-CREATE Training Grant is another example of a very prestigious training program that values interdisciplinary approaches and considers an association with the DEB graduate program a strong asset.

At the UC Davis 2006 Chancellor's Fall Conference on Graduate Education, the DEB was applauded as an innovative and successful graduate program that truly linked the PhD students to the real world. The DEB focus on translational research is very appealing to pre-doctoral students, particularly under-represented minorities (URM) and women. The number of under-represented and female DEB students has greatly increased over the last two years, reflecting this educational trend. We currently have 21 URM students: 14 of which are female. There were two URM students in the DEB in 2002. We added 1-3 students in years 2003-2005 for a total of 8. In 2006, we enrolled six. As of September 2007, we have seven new URM students, not including those that may enroll from the new crop of pre-doctoral students arriving on campus this fall.

Studies have shown that both under-represented minorities and women are drawn to areas of science, engineering and technology that focus on solving "real-world" problems.[3]

The DEB graduate program addresses the four principles stated in the report:

- Scholarship is the heart of the doctorate
- Concept of Service – seek ways to make the application of knowledge extend beyond the academy
- New People – the doctorate must be more socially responsive and less abstract, white, irrelevant
- New Partnerships – there must a continuous relationship between those who create the doctoral process and all those who employ its graduates

Last, but not least, the creation of start-up companies by DEB graduates are a testimony to the impact of the graduate program. Exposure to entrepreneurship during their doctoral training is having a positive effect on the students.

CAREER PATHS

About 50 percent of the graduates of the DEB program have chosen careers in the biotechnology industry. Four of our graduates were hired by the companies in which they interned. A couple of students who entered the UC Davis Little Bang and Big Bang

---

3   *Woodrow Wilson National Fellowship Foundation (September 2005)*
     "The Responsive Ph.D. Report."

Business Plan Competition have started their own companies. A few have completely left research. One graduate is attending law school at Boston College with the goal of becoming a patent attorney. She discovered her passion for the law while filing for a patent during her doctoral research in plant biology. The remainder of DEB graduates has entered academia as lecturers, faculty, post-doctoral and staff researchers, and as administrators.

The Biotechnology Program and the DEB graduate program address local economic development needs. Along Northern California's I-80 corridor, significant growth in the biomanufacturing sector of biotech industry is occurring, due in large part to the expansion of Genentech and Novartis. By 2010, more than 120,000 estimated AS/Certificate and BS biotechnician positions will be available in the California job market, an increase of ~40 percent since 2000.[4] In addition to entry level positions, there is an increased demand for qualified master's and doctoral researchers. In testimony to the California Assembly Select Committee hearing on Biotechnology in January 2006, Matthew Gardner, President of the Bay Area Bioscience Center (BayBio), reported an estimated 8000 new jobs were expected to be created in the life science industries in Northern California, over the course of one year.[5]

Given the regional significance of the biotech industry, as an employer and vital driver of economic growth, the Biotech Program strives to ensure that local populations are aware of educational opportunities and technical training programs in biotechnology. The region currently offers a biomanufacturing training program and biotechnology certificate through Solano Community College (SCC), and biotechnology degree and certificate programs at American River College in Sacramento (ARC).

Both UC Davis and the local community colleges have excel-

4   Peters, J. and Slotterbeck, S. (June 2004) "Under the Microscope: Biotechnology Jobs in California." California Employment Development Department.

5   Matthew M. Gardner, BayBio. (Jan 2006) Testimony to Assembly Select Committee Hearing on Biotechnology.

lent educational programs that emphasize hands-on training and soft-skills valued by the biotech industry. For example, the Solano Community College biotechnology program is taught by experts in the field having strong backgrounds in biotech industry and, specifically, experience in bioprocessing. Program lead instructors Dr. James DeKloe and Dr. Ed Re have established relationships with industry and work closely to align the college program with the industry skill standards for biomanufacturing. The biotechnology program at SCC confers approximately 25 degrees and certificates annually. To underscore the value of these technician training programs, many students that begin academic programs in biomanufacturing at SCC are recruited to industry before completing their academic coursework!

The DEB graduate program provides a valuable pool of talent to the region's biotech industry and, as previously mentioned, students in the program are highly sought after as interns and applicants for research positions. In addition, the Biotechnology Program takes an active role in supporting the expansion of local biotech industry in the region, placing three Chemical Engineering DEB interns to help Novartis AG in Vacaville gear up their process validation efforts for multiple products in the winter of 2007. Not only are chemical engineers in demand, but molecular biologists, biostatisticians, bioinformaticists and many other DEB subpopulations are popular with companies. Two local companies, Novozymes, Inc. and Monsanto, Calgene Campus take two-to-four DEB interns annually. Over the past few years, three DEB graduates have been hired by these local industry partners.

ADVANCED DEGREE PROGRAM (ADP) FOR CORPORATE EMPLOYEES

This is a unique program to retain outstanding employees in biotechnology companies. A BS/MS level employee can pursue a PhD in life science or engineering at UC Davis, while maintaining employment by their company. It is a win-win academic-industry partnership. UCD gets highly motivated graduate students with research experience and the corporation gets a loyal employee with increased skills.

Established in the mid–1990s, the ADP evolved as a companion-activity to the DEB internships. An industry executive suggested that UC Davis should enroll their industry employees, who wanted a PhD, since they were training our DEB graduate students. It was agreed that this proposal was a great idea. Prof. JaRue Manning, Section of Microbiology, became the first ADP director. Dr. Judy Kjelstrom took over as director in 2000, so the Biotechnology Program functions as the administrative home. Currently, the ADP is jointly run by the UC Davis Biotechnology Program, the College of Biological Sciences, and the College of Engineering, in cooperation with the Office of Graduate Studies.

The current list of PhD programs offered for ADP include: Biochemistry & Molecular Biology (BMB); Cell & Developmental Biology (CDB); Genetics (GGG); Molecular, Cellular & Integrative Physiology (MCIP) and Plant Biology (PLB). Chemical Engineering and Materials Science were added in 2005 so that the College of Engineering could participate. More graduate programs will be added as the demand increases. The ADP is still a fledgling program, since it is difficult to convince some companies of the value of investing in their employees. As the shortage of well trained technical personnel become more apparent, we predict a rise in industry participation.

The logistics of the ADP are:
- The corporate employee works with the ADP director to apply to a graduate program at UC Davis. The application packet is reviewed and finalized.
- ADP students spend 1-2 years on campus, completing their coursework, then returning to their company to conduct most of their research.
  - ADP students have both UCD and industry PhD mentors.
  - Research projects must not be tied to the IP of

the company, so that participating students may freely discuss their research in the academic milieu.

- ADP students add a level of focus and professionalism to the classroom leading to a cross-fertilization of ideas
- ADP student projects may catalyze future collaborations between the UCD faculty and the industry partner.
  - Example: Prof. X's collaborative research with Berlex on cardiovascular projects. He mentored a Berlex employee in the MCIP grad group, as well as sending a few UC Davis PhD students to Berlex.
- Participating ADP companies donate $5000 per year per student to UCD to support graduate education programs. A small portion is deducted for administrative costs in the Biotechnology Program
- There is an annual luncheon to honor the ADP graduate students and their corporate mentors. Many UC Davis faculty and administrators attend. Potential company partners are also invited, with the goal of raising corporate awareness of the ADP's value.

There have been two ADP students from Novozymes, Inc., located in Davis, California. One has graduated and was promoted, while the other is beginning his second year of pre-doctoral study. He is working on a biofuels project with a UCD plant biologist, highlighting the value of the ADP in encouraging collaborative research between UC Davis faculty and biotech companies.

SUMMER SHORT COURSES

The Biotechnology Program, in partnership with UC Davis's Genome Center Core Facilities and industry sponsors, offers one-week technical lecture/lab courses for graduate students, faculty, post-doctoral researchers, staff researchers and industry scientists. These courses serve an important training niche as new technologies emerge. Recently offered courses include topics in: *Flow Cytometry; Advanced PCR Techniques; Proteomics: Fundamentals & Technology Platforms; DNA Microarray Theory, Techniques & Analysis*; and *Bioinformatics*. The Biotechnology Program is also involved in "Train-the-Trainer" workshops for high school and community college teachers.

Flow Cytometry. Co-sponsored with FloCyte and the UC Davis School of Medicine
Lead instructors include: Carol Oxford, Manager, UC Davis Optical Biology Lab; Susan DeMaggio, President and CEO, FloCyte Associates, Inc.; Barbara L. Shacklett, PhD, Assistant Professor, Dept. of Medical Microbiology and Immunology, School of Medicine, UC Davis.
Flow cytometry is a powerful tool in modern biology. The applications are numerous, and have contributed to many fields, including virology, immunology, and molecular biology. This course begins with an introduction to flow cytometry and progresses through the development and design of multicolor panels. Afternoon sessions will include hands-on training in applications and training on various instruments. Students will be exposed to the Cytomation MoFlo cell sorter, Becton Dickinson FACScan flow cytometer, Coulter XL-MCL flow cytometer and Compucyte laser scanning cytometer.

Advanced PCR Techniques. Co-sponsored by the Lucy Whittier Molecular Core Facility.
Lead instructor (2003-2005): Dr. Christian Luteneggar, former Director of the Molecular Core Facility. The course has not been offered in 2006 or 2007 due to the lead instructor's movement into an industry position.
This very popular course covers: sampling strategies to obtain high quality biological material; micro-dissection of frozen or

paraffin embedded tissues using the new state-of-the-art PALM Laser Catapult (Zeiss, PALM Microlaser Technologies); sample preparation and nucleic acid recovery from tissue, blood and paraffin samples; quality controls for sample preparation, and real-time TaqMan PCR with contamination control and monitoring. Students can work on their own samples, including micro-dissection of frozen or paraffin embedded tissues, RNA extraction, QC and gene profiling.

Proteomics: Fundamentals & Technology Platform. Co-sponsored by the UCD Genome Center Proteomics Core

Lead Instructor: Dr. Brett Phinney, Director of the Proteomics Core, has taught similar proteomic courses at Cold Spring Harbor in New York.

This introductory proteome analysis workshop is designed to expose participants to fundamental technology platforms and current information in the field of proteomics. The course covers quantitative and non-quantitative liquid chromatography (LC) based proteomics and multidimensional protein identification technology (Mudpit) employing complex mixtures of proteins. Participants also perform quantitative ITRAQ experiments, in-gel digestions, liquid isoelectric focusing, and database searching using X! Tandem and Mascot with MS/MS data generated during the course. Lectures cover fundamentals of protein chemistry and mass spectrometry (MS), MS-based protein identification, post-translational modification of proteins, database search using MS data, and traditional tools of protein biochemistry. Lab sessions will include hands-on practical applications of the techniques, using state of the art instrumentation.

DNA Microarrays: Theory, Techniques and Analysis. Co-Sponsored by the DNA Microarray Core Facility and Affymetrix, Inc.

Lead Instructors: Satya Dandekar, PhD, UC Davis School of Medicine, Medical Microbiology & Immunology, and Director of the Microarray Core Facility, and Katrin Stapleton, PhD, Affymetrix, Inc.

This hands-on course is a must for any researcher wanting to explore the global analysis of gene expression through DNA microarray analysis. Lectures are presented by industry scientists, as well as academic researchers on the basics of chip technology, including analysis and quality control, and application of this technology in a wide range of research fields. The laboratory sessions will include preparation of cDNA from RNA, DNA hybridization to, and processing of, the microarray chip. Computer laboratory sessions will include data analysis with clustering, as well as other bioinformatics tools.

One Day Workshop in Bioinformatics - Browsing Genes and Genomes with Ensembl
Lead Instructor: Bert Overduin, PhD, Ensembl helpdesk officer at EBI
The UC Davis Bioinformatics Core at the Genome Center and the UC Davis Biotechnology Program presented this half day workshop primarily targeting laboratory researchers. The Ensembl project[6] provides a comprehensive and integrated source of annotation for vertebrate, genome sequences. This comprehensive workshop introduced participants to the Ensembl Genome browser, in addition to the BioMart data mining tool.

Train-the-Trainer Workshops

Train-the-Trainer Programs for Community College and/or High School Science Teachers have been run by the Biotechnology Program, in partnership with American River College, for many years. It is understood that we cannot address future workforce needs in healthcare, biotechnology and other life science disciplines, if K-14 teachers are not well prepared. Our current effort is a joint NSF-ATE project with American River College: *Applied Biotechnology and Bioinformatics for High School Teachers* (DUE-0603481; 2006-2009). Dr. Jamison-McClung is the lead instructor for the week-long workshop, with follow-up sessions throughout the school year.

The Biotechnology Program recently completed a similar joint NSF-ATE project with American River College: Functional Genomics and Bioinformatics for Community College Faculty

6   *http://www.ensembl.org*

(DUE-0302940; 2003-2006). Dr. Jamison-McClung was also the lead instructor for the week-long workshop. From 2000 to 2003, Kjelstrom coordinated the NSF-ATE grant: Tools to Teach Molecular Biology and Bioinformatics to Community College Teachers (DUE-0053291; 2000-2003). In the 1990s, the UC Davis Biotechnology Program led the way in training community college instructors and it is recognized that the majority of biotechnology programs in California Community Colleges spun out of these UC Davis workshops.

ENTREPRENEURSHIP ACTIVITIES

UC Davis, especially through the Office of Research's Innovation Access, is intimately involved in developing a life science, clean technology, and biotech cluster in the greater Sacramento Region. The UCD Biotechnology Program is an active supporter of entrepreneurial efforts on campus and in the region. Dr. Kjelstrom serves on the Board of Directors of the Sacramento Regional Technology Alliance (SARTA) and as a consultant for Velocity Ventures (a venture capital company focused on early stage technology companies in the region). The Program is an official partner of the Little Bang Poster Competition (led by Meg Arnold of the UCD Office of Research's Innovation Access), which feeds into the Big Bang Business Plan Competition (run by the Graduate School of Management's MBA students). The program also supports the activities of the Center for Entrepreneurship. PhD students in the DEB program have participated in all of these activities and have even launched their own companies.

The motivation for the UC Davis Biotechnology Program's involvement in all of these activities is to serve as a catalyst to expand the workforce in biotechnology, ranging from AS-degreed biomanufacturing technicians, trained at Solano Community College, to PhDs in science and engineering with a Designated Emphasis in Biotechnology and Certificate in Business Development, as well as support entrepreneurship in the region.

UC DAVIS BUSINESS DEVELOPMENT COMPETITIONS

The Little Bang Poster Competition[7] is an annual competition for students interested in moving their ideas to market. Teams are encouraged, but not required, and may be of any size. Individuals have participated quite successfully in the past, although in business, work is often accomplished in teams. Usually an MBA student joins a science or engineering team to create a research poster, identify market uses for the research and communicate both scientific findings and market opportunities. Graduate, post-doc, and undergraduate students are all encouraged to participate. Mentors from industry or venture capital firms are strongly encouraged, as they provide critical assistance with the entrepreneurial and business aspect of the competition. The Little Bang organizers help match teams with mentors.

Posters in the Little Bang competition combine research and technologies originating in UC Davis labs with entrepreneurial and business information such as: 1) How would the technology move into the market?; 2) What is the market opportunity -- how many people would need it, or want it, and at what price?; 3) Who are the customers for the technology, and what is the best way to market and sell to them?; 4) What steps are needed to turn the technology into a saleable product or service, and how long would it take? ; 5) How much money would a start-up company need to bring the technology to market, and where would that funding come from?

Cash prizes are awarded to the winners in each of five categories: 1) Medical & Biotech Innovations; 2) Nanotechnology; 3) Foods for Health and Wellness; 4) Clean Energy/Environmental Sciences and 5) Computational Sciences and Information Technology. The Little Bang is a great entry point into the Big Bang! Business Plan Competition. In 2007, four of the six finalists came from the Little Bang. It is also a great way to explore an interest in entrepreneurship and business.

The Big Bang![8] is the annual UC Davis Business Plan

7   http://littlebang.ucdavis.edu
8   http://bigbang.gsm.ucdavis.edu

Competition organized by MBA students of the Graduate School of Management. The goal of the contest is to promote entrepreneurship at UC Davis and throughout the region, supported by the University. Big Bang! provides a year-round forum in which UC Davis students, alumni, staff and faculty can collaborate to develop and test their business vision and plans. The competition provides a network of resources for mentorship, team creation, education, networking and financing for these aspiring entrepreneurs. A number of start-up companies have been launched through this competition.

CENTER FOR ENTREPRENEURSHIP

Both the Office of Research (OR) and the Graduate School of Management (GSM) are embracing the campus' strategic plan, including objectives to increase the University's impact in regional economic development. New initiatives and programs that enhance the level and quality of technology transfer, business development and industry interactions in both education and interdisciplinary research programs receive strong support. Furthermore, a substantial increase in the number of invention disclosures generated by the campus, coupled with the University's development of the Innovation Access unit within the Office of Research, has promoted a culture of entrepreneurialism among our researchers and, through the classroom, our students.

- The GSM, in collaboration with the Center for Entrepreneurship, offers a one-year program in Business Development for graduate students and postdoctoral fellows in science and engineering. Inaugurated in the Fall of 2004, the Business Development Program is directed by Prof. Andrew Hargadon. It provides training in a wide range of skills necessary to commercialize research, whether in new venture start-ups or in corporate research and development settings. These skills are intended to prepare scientists and engineers for careers in entrepreneurial firms, as well as industrial research and development.

- The *Business Development* courses offer both theoretical and hands-on training in developing new business ventures designed to commercialize research. Students take courses in technology management, innovation, and entrepreneurship. Hands-on training includes participation in practicums, working in interdisciplinary teams alongside GSM students and under the guidance of GSM faculty, Innovation Access staff, investors and entrepreneurs. Students completing the program are awarded a Business Development Certificate.
- The GSM also offers a one week boot camp in business development for scientists and engineers. Interested students can attend the week-long course before committing to the intensive year long program.

Over the past several years, a number of DEB faculty/students have participated in entrepreneurial activities on campus. Three DEB students have successfully completed the GSM's Business Development Certificate Program since 2004. In 2006, both the Big Bang Business Plan Competition and the Little Bang Poster Competition were won by a team co-captained by a DEB student. Six additional DEB students and a faculty trainer have place highly in the Big Bang and Little Bang competitions since 2003.

According to Andrew Hargadon, the surprising truth about how companies innovate is that "Revolutionary innovations do not result from flashes of brilliance by lone inventors or organizations."[9] Dr. Hargadon contends that Edison, Ford, Watson and Crick, Steve Jobs, and others were no smarter than the rest of

9   Hargadon, A. (2003) *How Breakthroughs Happen. Harvard Business School Press. Boston, Massachusetts.*

us—they were just better at moving through the networks of their time. He states that "Innovation is really about creatively recombining ideas, people and objects from past technologies in ways that spark new technological revolutions." Technology brokering is the key to bridging gaps in existing networks. The goal is to create dense social networks that enable a smooth flow of capital-human, intellectual, financial and social.

K-14 OUTREACH (FILLING THE PIPELINE)

The mission of the regional biotechnology consortium, BioTech SYSTEM, is to promote biotech training, educational and mentoring programs across Northern California. In the summer of 2005, this grass roots effort was launched by Dr. Judith Kjelstrom, Director of the UC Davis Biotechnology Program, Dr. Jim DeKloe of Solano Community College, Yen Verhoeven of Vacaville High School and the Solano County Office of Education ROP (Regional Occupational Program). The administrative home of the BioTech SYSTEM is in the UC Davis Biotechnology Program and is directed by Dr. Denneal Jamison-McClung.

The BioTech SYSTEM serves as an intermediary for members of biotech industry, academic training programs in biotechnology and members of local government responsible for ROP's and workforce development in biotechnology. All of these partners share a commitment to: 1) K-14 science education; 2) promoting biotechnology awareness; and 3) supporting technical workforce development in the region.

Activities of the BioTech SYSTEM include:
1. Co-organizing and fundraising for the annual Teen Biotech Challenge (TBC), a competition for high school students in the greater Northern California region
2. Creating and maintaining a consortium website[10] to serve as a portal for biotechnology education,

---

10 http://biotechsystem.ucdavis.edu/

training and mentoring opportunities

3. Developing a local, teacher-accessible Biotech Speaker's Bureau on various biotech-related topics, as well as career pathways

4. Co-sponsoring and organizing Biotech Career Day science fair events at area high schools and community colleges

5. Organizing teacher tours of local biotech companies and cutting-edge Research Centers at UC Davis (Center for Biophotonics Science & Technology; Center for Virtual Care; Genome Center)

6. Participating in the Explorit Science Center Challenge to raise funds for the Science Center in Davis, California (2006 Science Challenge table sponsored by Mars Inc.)

A critical element in K-14 outreach efforts is the training of high school biology teachers in molecular biology and bioinformatics. A new NSF Grant (ATE-DUE-0603481) was funded in 2006 to address this need. The title of the grant is "Applied Biotechnology and Bioinformatics Training for High School Teachers." The grant is a partnership with American River College and UC Davis.

The week-long teacher training occurs during the summer at UC Davis and involves lectures, hands-on computer lab exercises and curriculum development. Follow-up weekend workshops on more advanced topics, such as database construction and design of bioinformatics webpages, are scheduled for the academic year. To support participating teachers that do not have access to computer labs, a loaner set of laptop computers is available for check out. The program director will assist the teachers in implementing bioinformatics curriculum custom-designed during the summer workshop. Hopefully, these teachers will urge their students to enter the Teen Biotech Challenge (TBC)[11].

The BioTech SYSTEM, along with the North Valley & Mountain Biotechnology Center and the UC Davis Biotechnology

11   http://biotechsystem.ucdavis.edu/TBC

Program, developed the TBC, now in entering its fourth year. The TBC was launched in 2005 by American River College and the North Valley & Mountain Biotechnology Center, in partnership with the UCD Biotechnology Program. In 2006 and 2007, the BioTech SYSTEM took over the leadership of the annual contest. The TBC provides a vehicle to promote biotechnology awareness at the high schools and is currently very successful, with over 175 students from the Greater Sacramento region participating in the 3rd annual event. The TBC just keeps growing in the number of participants and the quality of submissions. This event is funded by donations from academia, industry, businesses and individuals who support science education. Participating students and teachers gain academic honors, cash prizes and the opportunity to interact with members of the biotech community, including those from industry.

The goals of the TBC are to:
1. Increase community familiarity with both science and computer technology
2. Target local high schools, rewarding high-achieving teens and teachers
3. Inspire teens to choose educational programs and career paths in science and technology.

THE TBC CONTEST

- High school students choose a topic from one of six focus areas in biotechnology:
  - Agricultural & industrial applications
  - Biofuels & bioenergy
  - Biomedical applications & bioengineering
  - Forensics
  - Genomics, proteomics & bioinformatics
  - Stem cells & tissue engineering
- Students design an educational webpage on their

topic, including an element of multimedia, such as an animation or video clip.

- Entries are evaluated based on the quality of webpage content and web design, overall creativity, organization and adherence to contest instructions. First- and second-place winners in each focus area are awarded cash and prizes, and a grand prize winner is selected from the first-place entries.
- Participants, parents, teachers and sponsors are honored during the Biotech Symposium Awards Banquet, held at UC Davis in May.
- Teachers of first place winners receive laboratory equipment and kits donated by industry.
- Winning webpages are compiled to create an Internet-based instructional resource for teachers, students and the community[12]

SUMMARY

The UC Davis Biotechnology Program, the Designated Emphasis in Biotechnology (DEB) graduate program, the Advanced Degree Program (ADP) for corporate employees, and the BioTech SYSTEM consortium could be replicated elsewhere, but it would take leadership from the university, as well as funding. Directors of these programs should be MS/PhD scientists who know the subject matter, but don't have to run research labs. The directors need to hone their managerial and networking skills, as well as raise funds through grants or donations. Social intelligence and project management skills are required in these types of programs. These programs need directors who can bring diverse groups of people together for a common purpose. Academic-industry-government partnerships are a must!

These programs have both an immediate impact and a long term impact on the community and the students. A number of PhD graduates with a designated emphasis in biotechnology have

12 *http://biotechsystem.ucdavis.edu/TBC*

been hired in regional companies as a result of their successful internships. A few have entered the Big Bang! and plan to start a company. Two employees from Novozymes, Inc., have participated in the ADP. An increasing number of our region's high school teachers are joining the BioTech SYSTEM and completing "Train-the Trainer" workshops. As a result, more high school biotech academies have been created and increased enrollments in high school biotech classes has been seen. Some of the excellent high school students emerging from these local programs are then able to complete summer research internships at UC Davis or local companies. Upon graduation, many of these students are pursuing science majors, some of whom choose Solano Community College, Sacramento State or UC Davis. Hopefully, the Greater Sacramento Valley will attract and sustain the growth of an increasing number of technology companies, keeping our technically skilled graduates in the region.

There has been success in promoting these ideas to business leaders, which is proving to be a critical factor in the continued expansion of biotech education programs in our region. Winnie Comstock's editorial on *Whose job is it?*, stressed the need for a technical workforce.[13] She also challenged the business community "to be more active at all levels of education, if they are to get the future workers they need...That involvement can, and will, take many forms: money; mentoring; internships; equipment loans and tangible connections to academic disciplines in your industry."

---

13 Comstock, W. (September 2007) *Whose job is it? Comstock's Business Magazine, Comstock Publishing Inc.*

University of New South Wales Diploma in Innovation Management
Wallace Bridge and Laurence Osen

Wallace Bridge, BSc, MAppSc, Ph.D., is Director of the Entrepreneurs in Science Unit at the University of New South Wales. Wallace can be contacted at *w.bridge@ unsw.edu.au*.

Laurence Osen, BSc, BA, DipInnovMan, is Assistant Director of the Entrepreneurs in Science Unit at the University of New South Wales. Laurence can be contacted at *l.osen@unsw.edu.au*.

The Diploma in Innovation Management is an undergraduate program offered by the Entrepreneurs in Science Unit (EIS) of the Faculty of Science, University of New South Wales (UNSW), Sydney, Australia.[1] The Diploma is open to UNSW undergraduate students commencing their second year of a four year science-based degree. Students are assessed for admission based on their past leadership and entrepreneurial activities, enthusiasm, and academic achievement. The program aims to encourage an entrepreneurial mind-set and provide students with the knowledge and skills needed for developing business opportunities based on scientific innovation. The program is primarily directed at empowering future R&D scientists with an entrepreneurial education that will allow them to recognise, evaluate, finance, and exploit commercial opportunities arising from scientific discovery.

The diploma is taught concurrently with the student's final three years of a BSc and involves the equivalent of 75 percent of a normal academic year divided into formal courses and an industry work placement. To avoid study overload, several of the program's core courses are delivered in winter or summer sessions. At the completion of 4 years of study, students become eligible

---

1   http://www.eis.unsw.edu.au

for the award of a combined BSc and the Diploma in Innovation Management (DipInnovMan). The industry work placement component is normally undertaken following the completion of the student's honours program or 4th year of university study.

Course content includes creativity in enterprises, project management, strategic communication, business principles, business planning, funding, marketing, accounting, finance, management and commercialisation of intellectual property, and valuation and assessment of high technology businesses. The program is focused on experiential learning to develop awareness and competency in commercialisation and entrepreneurship. Awareness comes from attending lectures and undertaking workshops and tutorials. Competence comes from real life projects such as starting up and managing a small business, event management, writing a grant application, writing a business plan, writing a scientific report for a peer-reviewed journal, business analysis of listed high tech companies, working in a high tech or commercialisation company or institution, and competing in virtual stock market games.

Prior to the establishment of the diploma, undergraduate BSc students at UNSW received high quality training in scientific fundamentals but little in the way of professional and business skills. Graduates would find themselves working in industry but have no understanding of business imperatives. The diploma was conceived to address this vacuum in the training of scientists by providing students with the knowledge and skills needed for developing business opportunities based on scientific innovation.

Key advantages of the add-on structure of the diploma are that students gain an extra tertiary qualification, and the science and technical content of their university studies are not diluted. Students studying in various fields across science, and engineering, are all eligible to undertake the program.

ENTREPRENEURS IN SCIENCE UNIT

The Entrepreneurs in Science (EIS) Unit was established in February 2001 within the Faculty of Science at UNSW. The mission of the unit was to establish a range of undergraduate and postgraduate innovation management study programs for science educated students. These programs would integrate business and entrepreneurial training into the core tertiary education of Australia's future scientists. The vision was to empower Australia's next generation of research scientists with a relevant education that will allow them to recognise, evaluate, finance, manage, and exploit commercial opportunities in their work.

The worldwide trend is for scientists to become business literate, not for business graduates to become science literate.

The diploma program integrates business and entrepreneurial training into the core tertiary education of undergraduate science students. Australia, like many other developed and developing nations, needs a critical mass of science graduates with business and entrepreneurial skills if it is to grow a successful local high technology industry which services global markets.[2] Science graduates must understand the path required for commercial exploitation and have the skills, knowledge, and networks required to contribute to this process. The expected outcomes of receiving such training are that our future scientists will be able to identify and effectively act on commercial opportunities derived from their research, and graduates will also be appropriately trained to work in the peripheral service career paths involved in the commercialisation of scientific discoveries.

THE NEED FOR ENTREPRENEURIAL SCIENTISTS

Science, engineering, and technology (SET) discoveries or developments are the cornerstone of the new global high technology economies. As we progress towards a global high technology marketplace and economy, more and more of the R&D advances and career opportunities will occur in industry or be sponsored by industry. Industry is in the business of making money, and to do so it requires a healthy pipeline of new products and technologies in various stages of development. To survive in business tomorrow

---

2   *Organisation for Economic Co-operation and Development (2006). Science, Technology and Industry Outlook 2006, OECD, Paris.*

a company must innovate today. To this end, industry requires R&D scientists who understand the innovation process and can use it to commercially focus their research efforts. Industry based scientists must understand business imperatives and how they can be achieved. Furthermore, scientists are becoming more and more recognised as being the most critical assets in high technology businesses—businesses which are totally dependent on the development and commercial exploitation of valuable intellectual property.

The Australian Prime Minister's Science, Engineering and Innovation Council (PMSEIC) 1999 report Ideas for Innovation strongly recommended the need for creating an entrepreneurial culture within these industries, and that the development of business skills by scientists and technologists should be a high priority.[3] Both federal and state governments have implemented schemes aimed at developing Australia's economy through enhancing business innovation, building from a strong foundation of research and development in science and technology. However, to achieve this highly desirable outcome will require science graduates that are able to bridge the gap between science and business, and whose minds are alert to the business opportunities that can be gleaned from scientific discovery.

Nations and cultures that nurture and encourage innovation and entrepreneurship in their society and back it up with a solid education system will be the real financial winners in the new economy. OECD reports have demonstrated that high investment in science and technology R&D by public and private sectors correlates with strong economic growth.[2] Innovation is now recognised by all major Australian political parties as being essential to the future strength of Australia's economy. The Australian public is demanding Government policies that will lead to profits from Australian know-how remaining in Australia.[4] All sectors of the

3   *Prime Minister's Science, Engineering and Innovation Council (1999). Ideas for Innovation, PMSEIC, Canberra.*

4   *House of Representatives Standing Committee on Science and Innovation (2006). Pathways to Technological Innovation, Canberra.*

Australian community believe that we must generate an entrepreneurial and innovative high technology culture to remain competitive in the global economy.[5]

For any nation to maintain and preferably progress its socio-economic well-being it needs to develop a substantial high technology industry with global markets. To succeed it needs to improve its ability to capture the value of its scientific discovery and engineering inventions. This will require a new generation of business savvy research scientists that can not only recognise commercial opportunity in their research findings but also know how to protect and commercially exploit their intellectual property.

In Australia, as is the case for most developed nations, a large proportion of new high technology based businesses are spawned from research activities conducted in government funded research agencies (GFRAs), such as universities, teaching hospitals, state departments and CSIRO (Commonwealth Scientific and Industrial Research Organisation). Most Australian universities have commercialisation or technology transfer administrative units which are responsible for identifying any potential intellectual property and for the development and implementation of strategies to gain a financial return on its eventual commercial exploitation. Academics and other government funded researchers are being asked by their institutions to look for commercial opportunities in their research findings and report them to these units for review. This leads to a situation whereby the majority of today's academics do not know how to adequately define the opportunity, let alone the markets. Representatives from the university commercialisation units, despite often being scientifically trained, do not adequately understand the technology. This mismatch will lead to good commercial opportunities being overlooked and other opportunities being pursued for much longer than necessary, where in reality they never had sound potential. As a consequence, risk is not reduced, resources are wasted, and true commercialisation

5   *Prime Minister's Science, Engineering and Innovation Council Working Group (2006). Australia's Science and Technology Priorities for Global Engagement [PMSEIC working group report], PMSEIC, Canberra.*

opportunities are lost.

A similar scenario exists for high technology start-ups seeking private equity investment. Again the business models are unlikely to be as sound as they might have been if the scientist was equipped with business acumen. The chances of unidentified or unmanaged technical, market and/or financial risk will be significantly higher. During the due diligence process, the investor may lack the required specific technical ability to identify some of this risk and as a result make an unwise investment. This will eventually lead to the loss of their investors' capital. Failure of the private equity-funded Australian high technology sector to deliver on the promise of high returns will see institutional investors turn away from early stage high technology venture capital funds with the firm belief that high risk equals no return.

Australia wishes to become a major player in the emerging high technology industries and to enjoy the inherent resulting economic benefits. High technology industries grow and develop from the seeds of scientific discovery. This means that for the good of Australia, all scientists are now charged with the somewhat alien responsibility of being commercially innovative. To be effectively commercially innovative requires a broad spectrum of business acumen, knowledge and drive. Most scientists are not innately commercially driven; otherwise they would have chosen business or finance as their career paths. Without a commercial mindset and the appropriate business training, it is unlikely that many scientists would be able to adequately identify and define any commercially viable opportunities that arise from their research efforts.

DIPLOMA STRUCTURE

At an undergraduate level, an obvious solution would be to include appropriate business focused subjects in the coursework; and some Australian universities are taking this approach. However, this strategy can potentially result in the science content of the coursework being diluted to an extent that the students do not receive sufficient training in the actual science. Though the graduates may be business educated they may not have sufficient scientific grounding to become research scientists capable of making cutting edge discoveries.

The Diploma in Innovation Management structure of housing the business related content in a separate add-on study program has several advantages. It does not involve dilution of the science content of normal BSc degrees, and it can be offered to all science students and not just those studying in a particular discipline; it does not compete with traditional science courses; and can be completed in the same time frame as the science degree providing a proportion of the business courses are taken outside of main academic sessions. By offering a separate non-imbedded qualification, the option for studying entrepreneurship and innovation management becomes open to all science students without losing flexibility in their choice of science subject study choices.

The major advantage of including the diploma training at an undergraduate level is the students are trained in business before they leave university and commence their careers. Current postgraduate options are more tailored for graduates that return to study when they recognise they need further skills to pursue their chosen careers.

Students that complete the diploma acquire skills which ultimately give them a greater range of career path opportunities and are more competitive for all employment opportunities. An added benefit for UNSW is that the diploma is progressively assisting it to be regarded as the best first option for high achiever local and international students intending to study undergraduate science.

Students cannot enroll in the diploma program without concurrent enrollment in a UNSW undergraduate science or engineering degree. They cannot graduate with the diploma unless they have fully met the requirements for graduation from their primary degree.

SCIENCE STUDENT MOTIVATION

The vast majority of undergraduate science students are not studying science so they can get an education that will allow them to create the next successful high technology company.[6] Most students study science simply because science is interesting and not because it offers great financially rewarding and commercially orientated career paths. Students that might have an interest or even passion for science but are commercially orientated are more likely to forsake their preference for science to study law, medicine, business, and other professional courses that have more clearly defined career paths that can offer a perceived higher standard of living. Some students may opt to undertake a combined degree of science and law or economics, but in most instances these two fields are taught totally independently of each other. Most graduates with combined science degrees tend to pursue careers relating to their non-science qualification, often simply because they offer the greatest employment opportunities.

The diploma structure offers a mode by which business minded students interested in science can have the best of both worlds.

TARGETED OUTCOME

The diploma content teaches students about, creativity, high technology drivers, economics, benchmarking, market research and analysis, IP management, finance and accounting, funding strategies, strategic communication, project and organisational management, business principles and business planning, business analysis, government regulation and policy, ethics (business and scientific), and legal issues and contracts.

Today's appropriately-trained scientist should be able to identify potential commercial opportunity in their research outcomes, whether they are working in industry or academia. They should know how to benchmark their technology; define and describe the associated intellectual property; identify and value potential markets for their technology; identify competitors and potential partners and customers; map the envisioned innovation and commercialisation process; address regulatory issues, prepare the ini-

6   Department of Education, Science and Training (2006). Science, Engineering and Technology Skills Audit Summary Report, DEST, Canberra.

tial business and project plans, budgets and timelines; strategically communicate; and, negotiate the advantages of the technology and the business proposition to a range of audiences. They should also have an enhanced capability of identifying and addressing, if possible, any potential risk, whether it is of technical, financial, or market origin. As a result, when the scientist contacts their commercialisation unit, the opportunity will be well defined and the decision as to whether to proceed with the commercialisation of the technology will be consequently of lower risk. Scientists will take on the initial task of screening and refining the opportunities and allow the limited resources of commercialisation units to actually focus on the commercialisation process rather than the screening and evaluation of opportunities.

ENROLMENTS

Students normally apply to enter the diploma program early in the second year of their undergraduate studies, with teaching commencing in the break midway through the academic year. By waiting one year before commencing the diploma, students are given the opportunity to adjust to university life and its responsibilities.

Selection for entry into the diploma is competitive, and is based in part on academic achievement and evidence of entrepreneurial interests and achievements, such as involvement in school leadership roles or in student societies, in business ventures or other indications of initiative.

The majority (70 percent) of the students study the bioscience disciplines of biotechnology and medical science. Approximately 15 percent study nanotechnology, with the others majoring in environmental science, chemistry, mathematics, and psychology.

A significant proportion (12 percent) of the students are taking combined BSc programs with law, commerce, or arts. The diploma program is structured to bridge the gap between these fields. The program complements and enhances these fields rather than competing with them.

In 2007, 164 UNSW undergraduate students were enrolled the

various stages of the diploma program. Annual enrolments have increased steadily since the diploma's inception, growing from 24 students in 2001 to 55 in 2007. The first cohort of students graduated in 2004.

MARKETING ISSUES

The marketing difficulty in attracting students to apply for entry into the diploma program relates to the fact that the majority of undergraduate science students are studying science for interest sake only and not for the commercial opportunities. Science students do not queue up to study business! If they were interested in business they would have enrolled in business in the first place. In many cases, their high school education has not reinforced the linkage between science and industry[6] and many, if not the majority of, current science students will not realise the absolute importance and relevance of business skills until they get out into the real world and find a job.

The diploma program is administered internally at UNSW and most students are not aware of the existence of the program prior to their enrolment at UNSW in their undergraduate science programs. A marketing strategy that has proven highly effective is sending personalised emails to eligible students. These "invitations" acknowledge the student's personal level of academic achievement, and discuss how the diploma program would be appropriate to the specific career opportunities afforded by their specific primary degree.

GRADUATE DESTINATIONS

Of the diploma program graduates, approximately one third tend to continue on with postgraduate science-based PhD programs, with many expected to ultimately find employment as research scientists in industry, universities and other government funded research agencies (GFRAs).

Diploma program graduates are proving themselves to be highly competitive for a broad range of career and employment opportunities. They are being employed in laboratory based positions (20 percent) or are pursuing other careers and training (50 percent) requiring a tertiary science background.

Combinations of business and technical skills are required in careers that involve intellectual property, high technology finance (venture capital) and investment (business analysis), R&D management, corporate management in the biotechnology industry sector; government regulations. Diploma graduates are employed in IP management (patent examiner/patent attorney), high technology finance (investment banking, consultancy, and business analysis), scientific communication (scientific journal editor), government policy, product development, high technology sales and marketing, clinical trial management, and business management.

Graduates can also be expected to fast track their careers, as they will be well prepared for more senior positions right from the start.

PROGRAM VALIDATION

The diploma program was new to UNSW and the first of its kind not only in Australia but possibly the world. As such, for the growth and reputation of the program it was essential that the diploma program be critically evaluated and ultimately validated by the key stake holders, the university community, the students, employers, and the high technology sector.

The extensive use of external partners including professionals from industry, government, and academia for the delivery of the diploma program offers significant validation of the program, as they gain some ownership. They teach it, and therefore consider the content appropriate. The partners also know who else is contributing to the teaching, they respect them and therefore, the broader content is deemed appropriate.

Diploma program students enter into a range of business based competitions, including the Young Achievement Australia (YAA) Business Skills Program and the Australian Graduate School of Management's (AGSM) Business Planning competition. The AGSM is ranked as South-East Asia's leading postgraduate business school. Competing against MBA students from Sydney metropolitan universities, Diploma program students have won first prize four times since 2003. Since 2001, diploma program students have received numerous YAA nominations for awards and have won many of them including NSW Tertiary Business Person of the Year (twice).

The work placements allow potential employers to get a taste of what a diploma program education offers. They are generally impressed with the student's proactiveness during the application process. Feedback on the student's performance against criteria such as; work ethics, time management, enthusiasm, interest, technical competence, communication skills and overall quality is, without exception, highly positive. Many students have been offered permanent positions following their graduation.

The EIS unit is regularly contacted by employers who have either had a diploma program student for a work placement or have hired a diploma program graduate to see whether other graduates that would be interested in a job.

TEACHING STRATEGY

The commercialisation path starts with an idea (or potential innovation). That idea may come from a discovery in a fundamental study (science first, opportunity second) or it may come from an identified market opportunity (opportunity first, science second). But ideas are a "dime a dozen", and it can be said that almost anyone can come up with at least one good idea in their lifetime. Commercialisation is not about the idea *per se*. Rather, it is about driving the idea into reality. Hence commercialisation is termed a process. The person who drives the commercialisation process is the entrepreneur, and the skills they need to drive it are defined by entrepreneurship.

Appropriate training for learning a process must be process orientated. That is, the students must go through the process to learn the process. Effective process teaching usually involves workshop style formats and small group projects.

The majority of undergraduate science students are not innately interested in business and many of them are extremely afraid of it. The program delivery style has to break down that fear, and engage and nurture the science student in order to ultimately generate enthusiasm for business (of course, at the same time you do not wish to damage that initial interest in science). A different style of teaching business to science students is required as opposed to what might be used for business students. Another issue to consider in experiential teaching is to make it as real as possible, in order to further motivate these students.

TEACHERS

The teacher should be experienced, if not an expert, in the field of commercialisation and entrepreneurship. For programs such as the Diploma in Innovation Management, the program authority would ideally be actively conducting scientific research in their field of interest, identifying potential market opportunities for their discoveries and attempting to commercialise them. This first-hand experience can then be funnelled into the education program's design, content, and teaching strategies.

ABSENCE OF APPROPRIATE EXPERTISE IN ACADEMIA

Most universities do not have a large pool of academic resources that are qualified to teach the appropriate material for commercialising high technologies. Science faculties tend to be staffed by academics that are experts in their particular scientific fields but on the whole have little training or expertise in business. Similarly, business and economic faculties tend to be staffed by academics with expertise in traditional finance, business and economics but they have little training and knowledge of science and the commercialisation of science.

WHERE THE HIGH TECHNOLOGY COMMERCIALISATION EXPERTISE LIES

High technology commercialisation expertise lies within the commercialisation arms of universities, government departments, high technology companies, service providers that supply specialised skills to the industry (such as IP management, accounting, finance, R&D consultants), professional associations, and technology parks (incubators). Many universities that are currently attempting to teach in the areas of commercialisation and entrepreneurship rely heavily on this extensive pool of expertise to teach their students. These professionals tend to be keen to donate their time to educate students in their specific area of expertise. They are keen for a number of reasons: they would like to promote their organisation to the next generation (potential customers or employees), and they often truly believe in the educational objectives and want to help ensure that they are achieved.

The EIS Unit has constructed a strong network of over fifty senior industry and government professional partners who routinely assist with the teaching of the diploma. The areas that these professionals contribute to include: program and course development; small group teaching (including lecturing and workshop facilitation); student mentoring; case study material; and the provision of industry work placements. Collaborating organisations also often provide monetary prizes for high achieving students and financially support student-run events focused around building a community within the program. The partners work collaboratively on topic areas to avoid duplication and to provide continuity in the program's content.

THE DYNAMIC CONTENT OF COMMERCIALISATION AND ENTREPRENEURSHIP

Some commercialisation and entrepreneurship skills involve fundamental rules and as such are static in nature. However, the majority are dynamic and change with variations in economies, technology, government regulations, policy, and industry trends. It is this dynamic content that should be taught by experts who are currently working professionally in the topic area.

Having a large number of external professionals contribute to the delivery of the diploma program content provides many advantages. Material is constantly updated and is therefore relevant. Students begin their professional networks early and all players in the sector gain some ownership of the program and hence validate it. Potential employers get to see potential employees (and vice versa), and the students gain insight into the real world high technology sector. Students are also able to see (and to experience through work placements) the myriad career opportunities that are open to them following graduation.

Issues to be addressed when running a program taught by fifty or more individuals include continuity, repetitiveness, context, and conflict of content. These potential problems can be readily addressed if the program authority does not merely ask a guest to come and give a presentation on what they choose. Rather, guest presenters should receive detailed guidelines on the requested content and the material being covered by other contributors to the course. Other material offered by the presenter is generally appropriate for inclusion and will be approved if it is considered by the program authority to be relevant to the course's objectives. Guests are also encouraged to use similar (preferably the same) examples from one presentation to another. For example, the same company will be discussed as a core case study throughout a particular diploma course.

It is essential that the program authority chair all presentations to ensure context and continuity are maintained throughout each course. Guest lecturers are never asked to become involved in the student assessment process (except for work placements). This should be the sole responsibility of the course coordinator, who must be completely familiar with all course content. As such, they should be in the academic position to assess all covered material and judge which material is the most appropriate to assess and test.

ALTERNATIVE OPTIONS TO THE DIPLOMA MODEL

Commercialisation forums offered by universities, government departments, and professional bodies are usually structured as short seminar series and occasionally include workshops. These are only sufficient at best to introduce the student to the key sections of the innovation process. Students are not actually trained in the various phases of the process but do have the opportunity to get a reasonable picture of what is involved. However, science students tend to not take advantage of these forums. This is because the majority, as stated previously, are not business focused and have yet to recognise the importance of business skills for their future careers. Secondly, most students are reluctant to take on extra studies if they do not receive any formal accreditation.

COURSES

Some diploma program courses are taught in one week intensive blocks at the beginning or end of summer and winter terms. The positioning of the out of session courses is to reduce the impact on the students' ability to take holidays and to engage in casual employment.

In order to allow flexibility for the students, who come from a diverse range of primary study plans, the majority of the Diploma courses are stand alone and do not have prerequisite or corequisite requirements.

CORE DIPLOMA IN INNOVATION MANAGEMENT COURSES

1. Introductory course: The Innovation Process
Structure: 5 day block: 3 x 1 hour morning presentations plus 4 hour afternoon workshop.
Assessment: Exam (50%), Workshop participation (20%), Individual Assignment: Preparation of a Professional Skills portfolio (30%)

This course gives students a fundamental and practical introduction to the innovation process from the creation/generation of the idea through to the preparation of a rudimentary business plan. It is not intended to get too deep into the details of each phase, but rather simply cover each section in terms of what is required and why it is important. Completion of this course allows students to develop a holistic perspective of the relevance of each phase and aspect of the innovation process and prepare them well for material to be presented in subsequent courses.

The lecture topics include: an overview of the innovation process from idea to market, skills of a high technology entrepreneur, action planning to determine technical or commercial merit, benchmarking and competitor analysis, market evaluation, strategic business planning, introduction to intellectual property (IP), technical evaluations and risk management, and funding sources. The course also uses case studies (by CEO's) to supplement the lecture content.

The overall objective of the workshops is to develop and evaluate at least two products that would potentially form the basis of a real small business that they would start up in the following year. The week starts with some theories on creativity and an introduction to a number of brainstorming activities. Through the brainstorming process, the class generates a number of product concepts that the class collectively feels have potential for success. Throughout the remainder of the week the class (divided into teams) refines and evaluates these concepts for their market merit and business potential. They prepare action plans, act on these plans, conduct market research, prepare a SWOT analysis, and attempt to address arising issues and key assumptions. On the last day each team of students deliver a "pitch" to the rest of the class on their concept. The class then votes to determine the two best product candidates.

2. Business Planning and Business Start-up Skills
Structure: Year-long team project
Assessment: Each company (team) is graded by the course coordinator (mentor) against defined performance criteria. This mark is distributed to individuals by a formal evidential student peer review process.

Student companies are guided through all the stages of a concentrated business cycle, including selling shares to raise capital, establishing the company organisational structure, electing an executive management team, researching, designing and producing goods or services to fill a profitable market niche in the community. The companies plan, develop and implement quality systems in the key management areas of finance, manufacturing, human resources, sales and marketing. Each company is required to prepare a business plan and annual report, and at the completion of the program liquidate the company and distribute dividends to shareholders.

Throughout the year, students tackle typical issues and challenges which confront commercial operations and develop skills for decision making, negotiation, creativity, communication, teamwork and networking, leadership, responsibility and accountability, and financial management and planning.

There are no formal lectures or tutorials as the course format is based solely on the formation and management of a company. Meetings take place in the form of weekly company conferences, board meetings and regular business activities. Each company is closely supervised by academic mentors who attend and monitor all company meetings.

3. Strategic Communication
Structure: 5 day workshop
Assessment: Exam (50%), Workshop participation (50%)

Workshops focus on developing communication effectiveness in a science-related business environment. This includes communication between individuals; between people with different specialties; between women and men; in negotiation; in emotionally-charged situations; in meetings; in networks; and about new ideas. The course gives students a practical introduction to the broad range of communication skills required by the successful entrepreneurial research scientist.

The course material provides a theoretical and practical instruction in evaluating, communicating and marketing technical information, ideas and opportunities to a variety of audiences. Workshops focus on the development of the student's interpersonal skills including oral presentations, persuasion, negotiation, networking, business ethics and leadership. Other areas covered include analysis of personality and audience types, risk perception, locus of control, negotiation of expert status, effective listening, enquiry and feedback strategies, meeting facilitation, and the preparation of media releases.

4. The Bioentrepreneurial Process
Structure: 5 day block: 3 x 1 hour morning presentations plus 4 hour afternoon workshop.
Assessment: Exam (35%), Workshop participation (10%), Assignment (55%)

Students gain a fundamental and practical training in the preparation, use and interpretation of business information. Material covered includes, accounting (double entry and software), financial statements, business valuations, stock markets, tax, contracts, economic and business principles with a focus on the special considerations and parameters particular to the entrepreneurial process involved in the establishment of high technology based businesses.

The individual assignment to be completed over the summer break involves the preparation of a detailed financial analysis of a student selected company benchmarked against a prominent Australian listed biotechnology company.

5. Commercial Biotechnology
Structure: Main academic session (4 hours per week, lectures, workshops and tutorials)
Assessment: Assignments (45%), Exams (55%)

The course lecture material focuses on the forms and processes of protecting intellectual property (specifically patents), structure of the biotechnology and related high technology sectors, capital raising, benchmarking, commercialisation strategies, ethics, regulations, GMP, and clinical trials.

The course assignments reinforce and build on the lecture material. Student teams select an ASX (Australian Stock Exchange) or NASDAQ listed company which is developing products in an area of the team's mutual interest (e.g. gene therapy, transgenics, etc). The teams are charged with critically reviewing their company in terms of IP (patent portfolio), management team, funding, strategic alliances, product pipeline, productisation, regulatory issues, target market, and benchmarking current and future competitors. Students also increase their knowledge of the biotechnology industry by competing in a Virtual Stock Exchange game, where they trade only in biotech stocks. They must evaluate market trends, and source announcements (e.g. patent granted, clinical trial success) and advice in order to justify each trade.

6. Professional Issues in Biotechnology
Structure: Main academic session (4 hours per week, lectures, workshops and tutorials)
Assessment: Assignments (65%), Exams (35%)

Lecture material covers the various sources and processes of funding for research and commercialisation (government and other grants, and equity investment), project and risk management, financial management, market evaluations, business valuations, business planning, commercialisation strategies, organisational theory, and due diligence.

The best way to gain a well rounded understanding of overall business imperatives in a biotechnology company is to write a business plan for a biotechnology start-up company. Student teams are required to identify a real or theoretical commercial biotechnology based opportunity sourced either from their own research efforts/areas, from a UNSW or external academic or researcher, or from an existing company. They are then required to prepare an appropriate government grant application for funding of further R&D and a business plan aimed at attracting equity investment to start-up a biotech business based on the opportunity. Through this process the students become familiar with all the considerations required to successfully operate in the biotech industry.

7. Work placements
Structure: 4-12 weeks full time in a company/position of the student's choice
Assessment: Written report and supervisor's evaluation

Diploma students are required to complete a minimum four week full time equivalent work experience program. The students must develop value propositions for prospective host organizations and must negotiate their own placements. Students are mentored through this process by academic staff. Preparation of an assessable report is required at the end of the placement period. Normally, the placements are taken following the completion of the student's formal coursework studies. Any organisation that is operating in a sector that is of interest to the student's future career path is usually considered suitable. The positions do not have to be paid and, as the placement is formally part of the student's university program, they are covered by the university's personal injury insurance policies.

The assessable written report is primarily focused on a critique of the host organisation rather than a review of the student's specific work placement tasks.

LOOKING TO THE FUTURE

The EIS Unit at UNSW has been actively involved in commercialisation and entrepreneurship education of undergraduate science students since its establishment in 2001. Annual enrolments in the Diploma in Innovation Management have increased from 24 students in 2001 to 55 in 2007. This trend confirms the progressively increasing demand from undergraduate science students for tertiary commercialisation and entrepreneurship training. The first cohort of diploma program students graduated in 2004. The program structure and content is continuously upgraded to utilise and address feedback gathered from students, graduates, and industry and government partners. This ensures that the program's quality and content is constantly improving and moving towards satisfying the desired outcomes of all stakeholders.

Graduates have proven to be highly competitive for all jobs that fall within, or service, the Australian high technology industry. Many have ended up in positions directly from university that were previously only suited to graduates backed with significant work experience. Employers regularly approach the EIS Unit hoping to employ a diploma program graduate, confirming that the diploma program's content is meeting the high technology sector's need for business savvy scientists.

The EIS Unit has sought national and international collaborations with industry, government and academic organisations to support its activities. Moreover, many of these organisations have volunteered their services to the EIS Unit without being approached. The unit has an ever expanding network of professionals willing to contribute to the content and delivery of its programs as they see their inherent value to the high technology sector and Australia's future. In consultation with the EIS Unit, Ngee Ann Polytechnic, Singapore, has established a study program (Certificate in Innovation and Enterprise) by which its graduates can articulate to the combined UNSW BSc/DipInnovMan program. This collaboration process is currently being repeated with another Singaporean Polytechnic college.

The add-on structure of the diploma program has been shown to have several key advantages. Diploma program graduates are equipped with commercialisation and entrepreneurship training without diminishing their depth of knowledge in the core sciences. Regardless of their chosen field of undergraduate science or engineering study, students are able to complete both a BSc (or BE) and the diploma program in the same time-frame. Graduates are also able to show employers that they have completed an additional tertiary program, which automatically sets them apart from other BSc graduates.

More recently, in 2007, the Australian federal government has allocated significant funding for similar programs aimed at postgraduate PhD science students. This funding, set up under the banner of the Commercialisation Training Scheme (CTS), has seen many Australian universities adopt an add-on model similar to that of the diploma program.[7] These programs run concurrently

---

7   *Department of Education, Science and Training (2006).*
    Commercialisation Training Scheme Final Paper (June 2006), DEST,
    Canberra.

with postgraduate research (PhD) programs and provide students with an extra tertiary qualification on graduation. The EIS Unit at UNSW has recently established such a program, the Graduate Certificate in Research Management and Commercialisation. These CTS programs (including the UNSW program) have similar objectives and economic imperatives as that of the diploma program, reconfirming that Australia sees this commercialisation and entrepreneurship training for scientists as a necessity, not just a luxury.

The University of Calgary Master of Biomedical Technology Program

Melissa Belluz, Kent Hecker, Derrick Rancourt, and Wendy Hutchins

Melissa Belluz, is a student in the Master of Biomedical Technology program at the University of Calgary, Canada. Melissa may be contacted at *mabelluz@ucalgary.ca.*[1]

Kent Hecker, Ph.D., is Director of Admissions, Faculty of Veterinary Medicine at the University of Calgary. Kent can be contacted at *kghecker@ucalgary.ca.*

Derrick Rancourt , Ph.D., is Graduate Program Coordinator of the Master of Biomedical Technology program at the University of Calgary. Derrick can be contacted at *rancourt@ucalgary.ca.*

Wendy Hutchins, MLT, Ph.D., is Microbiology Core coordinator and Biotechnology Training Centre Director in the Master of Biomedical Technology program at the University of Calgary. Wendy can be contacted at *hutchins@ucalgary.ca.*[2]

The biotechnology sector has been one of the fastest growing and research intensive industries since the 1970s, expanding from the food processing and agriculture industries to the fields of drug development, vaccines, medical diagnostics, agriculture, fuel production and forestry, leading to potentially unlimited applications.[3] As a result of this diversity, there is an identified need for trained personnel in areas of the industry including research and development, manufacturing, commercialization, marketing and project management. In response, post secondary institutions have created programs integrating science and business skills specific to the biotechnology sector.

The Master of Biomedical Technology (MBT) program was

---

1 *Submitted in conjunction with the practicum requirements of the Master of Biomedical Technology degree.*
2 *Corresponding author*
3 *Paul, J., & Lova F. (2005) Biotechnology-The New Age "Global" Industry. Global Business Review 6:315-321.*

created in 2000 at the University of Calgary, Canada in response to the growing need for highly qualified personnel in the field of biomedical technology for the small but growing local biotech industry. Originally, the MBT program incorporated new areas of bioinformatics, genomics, and proteomics with the previously developed courses in medical sciences to create a program that was aimed at the research and development (R&D) science area. The MBT program has evolved from its inception to best meet the needs of its students and the local biotechnology industry. The program now provides graduates with a wider range of competencies useful to the local biotech companies for business development and market analysis. As a result, MBT graduates now play a substantial role in the growth of this industry sector in Alberta.

THE MBT PROGRAM AND THE ALBERTA BIOTECHNOLOGY INDUSTRY

At the time of its creation, the MBT was one of only two graduate level programs in Canada that combined business and science course requirements. The Alberta provincial government supports the need for biotechnology as part of its life sciences strategy; a strategy aimed at economic diversification. Most biotechnology companies in Alberta are small innovative start-up companies. They need highly skilled personnel who have acquired the scientific and business skills necessary to be adaptive and flexible to the demands within a small biotechnology company. MBT graduates possess the ability to work both science-related as well as business-related roles in these companies. Providing the knowledge, skills and attitudes necessary for success in the Alberta biotechnology industry is a strength of the MBT program, which is reflected by its numerous graduates currently employed in the local Alberta biotechnology sector.

The MBT program prides itself on its connection with industry. Through its business lecturers, business mentoring program, networking events, and industry practicums, corporate involvement is infused throughout the program. This provides students perceptions, experiences, and contacts to which they otherwise may not have been exposed. In addition, this crucial relationship with the local biotechnology industry allows the program to tailor its objectives to the needs of the industry. The program's multi-disciplinary approach exposes the students to the critical areas of the biotechnology community including business development, project management, and intellectual property. The success of this relationship is reflected in the large number of local practicum placements and permanent or internship positions offered to graduates. More recently, it also resulted in the formation of scholarship program for incoming students (established in 2007).

MBT ADMISSIONS

Admission to the MBT program is competitive and currently capped at approximately 20 students per year in order to promote and maintain an intimate learning and training environment. The minimum requirements for incoming MBT students are second year-level cell biology and genetics, undergraduate biochemistry, and two of microbiology, immunology, physiology, and pharmacology courses at the senior level. Half of the students in the program have previous work experience of one year or more and are looking for a new direction in their career. Often, those who enter straight from their undergraduate degree are not keen on laboratory-based careers and are seeking broader career paths or a field in which they develop a passion to continue on to the PhD level. Part-time enrolment in the program is available for students with full time employment. In addition, students accepted to the University of Calgary Haskayne School of Business may choose to do a combination degree with either the Master of Business Administration (MBA) or the MBT portion in their first year. This program requires just over 2 years for completion regardless of doing MBA or MBT first.

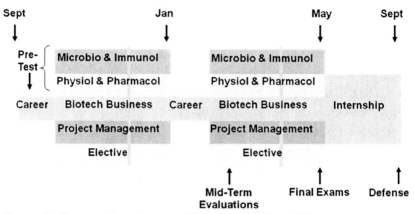

Figure 1: Course structure and timeline of the MBT program.

MBT CURRICULUM

The MBT program is a one-year course-based master's program whose mission is to provide education and practical experience in medical science disciplines related to biomedical technology and communication and business skills required for successful careers in biotechnology. This is accomplished through a series of core courses in physiology & pharmacology, microbiology & immunology, business aspects, and biotech career skills. Assessment of the integration of knowledge, procedural skills, and core competencies occurs throughout the program and in a final core examination.

Knowledge of biotechnology R&D methods is critical for all graduates, regardless of their desire for a lab-related or strictly business career in biotechnology. To facilitate training in laboratory theory and practice in the program, a laboratory was created—the Biotechnology Training Center (BTC)—which houses state-of-the-art equipment to educate MBT students in biotechnology-related techniques such as molecular biology, cell culture, and protein biochemistry. All MBT students have to attend the laboratory sessions, lab theory, and research design discussions, and a techniques journal club.

Students have the option of choosing a lab- or business-based project to work on throughout the year in order to learn aspects of project management, teamwork and individual leadership. They must attend monthly meetings, provide project charters, and progress reports about their individual efforts and team components. Student roles in the business projects may include the operations or business development of one of the MBT-owned/operated companies (e.g., Chief Executive Officer). These businesses allow MBT students to get exposure to aspects such as finance, regulatory affairs, managing growth and development, manufacturing, customer service, pricing, sales and marketing, corporate organization, and shareholders meetings. This project management course has led to many innovative projects, building on the creativity and strong work ethic of the MBT students. For example, Pipette Calibration Services and Sales International (PCSS) was created in 2003 by MBT students in collaboration with the BTC, which houses the company. To date, the company has remained self sufficient with numerous experienced and well trained student staff members who service over 1000 pipettes annually. The success of the company has lead to the creation of successful satellite ventures by MBT alumni and associates in Nigeria and Ethiopia, and has financed several student travel bursaries. Lab-based projects usually involve the development and validation of new biotech assays or diagnostics such as mycoplasma testing for cell culture.

MBT students are also required to take two elective courses from a wide variety of university approved courses, including bioinformatics, genomics, biology of laboratory animals, biotech and the law, health economics, determinants of health, bioengineering materials, or a directed study on a topic of the student's choice. In collaboration with the Haskayne School of Business, MBT students can also take any of the Entrepreneurship and Innovation courses offered through that faculty. This wide range of electives allows each student to tailor their MBT program experience to their personal interests and professional goals. To conclude the program the student has to do a ten to twelve-week practicum followed by an oral final exam and defence. The students conduct these industry- or academia-based practica/internships in various biotechnology realms including lab-based target validation, new diagnostics development, clinical trials or biotechnology-related business aspects such as patents, regulatory affairs, commercialization and market analysis. Figure 1 and Table 1 show the program schedule based on a full time enrolment in the MBT program. Part-time programs are tailored to the student's needs.

Table 1: MBT Program Requirements	
The program requirements consist of seven required courses, two optional courses, and the practicum	
Biotechnology Business Aspects: Aspects involved in taking an original scientific idea or discovery all the way to a start-up company are be covered. Lecturers discuss commercialization, venture capital, business plan, patents and law and marketing. Team projects include dissecting a mid-cap company from R&D, patents, marketing and financial perspective in addition to company case studies.	½ credit over Fall and Winter 1.5 to 3 lecture hours / week
Project in Biomedical Technology: Conduct a business- or laboratory-based project throughout the year. Laboratory components integrate with the core medical science course where possible and provide instruction in regulatory compliance in laboratory work and the use of laboratory assays in industry settings. Business projects as described above.	½ credit over 2 semesters 1.5 to 3 lecture hours / week 6 lab hours / week
Biomedical Technology Careers: A series of sessions designed to provide students with practical knowledge in general aspects of the biotechnology industry, such as career skills, public speaking, networking, career planning and professional development, and bioethics. Business mentors are required.	½ credit over Fall and Winter 2x1 week plus retreats
Integrated Systems Courses: **Principles of Physiology and Pharmacology** **Principles of Microbiology and Immunology** The principles of physiology, pharmacology, microbiology, immunology and pathology are taught from a practical perspective: working on development of new diagnostics, vaccines or therapeutics. Lectures in the two courses are in parallel and fully integrated. The goal of the courses, with an emphasis on molecular mechanisms in health and disease, is to provide students with the skills to interface with individuals in these disciplines in the biotechnology industry and to recognize novel ideas that could lead to new diagnostics, vaccines or therapeutic strategies. This is complemented by the laboratory exercises in the project course, demonstrations, tours and special lectures that provide industry perspectives in these disciplines. Journal club weekly.	4 credits over Fall and Winter 6 lecture hours / week 1.5 journal club hours / week
Practicum in Biomedical Technology: A laboratory-based or business-based practical experience carried out in an academic or industrial setting for a period of at least ten weeks. All projects are vetted for academic suitability to the MBT program.	2x ½ credits 10 to 12 weeks full time over Summer & Spring
Optional courses: Two to be taken over Fall and Winter or 1 in Spring	2x ½ credits

STUDENT ASSESSMENT

The MBT program's educational objectives were created using Bloom's taxonomy of knowledge, skills, and attitudes.[4] However, the program faculty felt this did not take into account the work skills of the mature students and was misleading to those direct from undergraduate programs who were trained on a knowledge-based system. This year, the program has been modified to use a competency-based model where attributes of a successful gradu-ate have been defined and measurements for performance have been designed. This is expected to better integrate the learning experiences and activities that have already been developed and the desired program outcomes. Numerous studies identified com-munication, teamwork, problem solving, and leadership as essen-tial generic competencies employers expect from post secondary graduates.[5,6] Studies also report that post secondary graduates are expected to be flexible, adaptive and adaptable and, therefore, general competencies should be incorporated into programs with specific competencies linked to course specific outcomes.[7]

By taking into account the needs of the biotechnology in-dustry in terms of employee competencies, the MBT program has strived to build learning experiences that will enable students to develop these necessary competencies and provide them with the essential skills, knowledge, and attitudes necessary for success in the biotechnology industry. Starting with the class of 2008, the students will be assessed three times throughout the program using the competency-based model. The first assessment at the

---

4   Bloom, B. (1956) Taxonomy of Educational Objectives: The Classification of Educational Goals; pp. 201-207; B. S. Bloom (Ed.) Susan Fauer Company, Inc. 1956.

5   Candy, P.,& Crebert, R. (1991) Ivory tower to concrete jungle. Journal of Higher Education, 62, 570-592.

6   Marginson, S. (1993) Arts, Science and Work: Work-related skills and the generalist courses in higher education. Canberra: AGPS.

7   Harvey, L., Geall V., & Moon S. (1997) Graduates' work: implications of organizational change for the development of student attributes. Industry and Higher Education, 11 (5), 287-296.

beginning of the program will be a student self-assessment. The students will assess themselves in the areas of personal effectiveness including confidence, organization, critical thinking, and analytic thinking. For interpersonal skills, the assessment will include communication, teamwork, leadership and networking. Students will

MBT 360 Evaluation Form

Name of Assessee: _____

Name of Assessor: _____

Scale	Beginner			Intermediate				Proficient		
	1	2	3	4	5	6	7	8	9	10
Confidence: Believes in oneself and one's ability.										
Independence: Works effectively without supervision.										
Organization: Maintains order and plans out work.										
Critical Thinking: Evaluates and applies information.										
Creativity: Combines elements to create new knowledge.										
Judgment: Assesses situations and draws sound conclusions.										
Involvement: Active participation in the program.										
Communication: Expresses ideas in a convincing manner.										
Teamwork: Facilitates cooperation within groups.										
Leadership: Influences others toward goal accomplishments.										
Comments or Examples:										

Figure 2a: MBT 360 Evaluation Form

also complete a goal-setting personal development plan, in which they outline their goals for a specific time period, challenges and areas of improvement. Students will be assessed midway and at the conclusion of the program. This will consist of a 360° evaluation (Figure 2) based on self-assessment, and assessment by peers, the program supervisor, and selected MBT faculty members. The feedback generated from these assessments should provide students perspectives into their areas of strengths, weakness and their overall progress throughout the program.

MBT FINAL CORE EXAMINATION: A NOVEL EXAM FORMAT

The MBT program has devised a comprehensive final core examination process to assess the ability of students to integrate the knowledge and skills acquired throughout the program. Unlike traditional final examination processes which are generally based on knowledge and comprehension, the MBT final core examina-

DEFINITIONS

Personal Effectiveness
Confidence: Believes in oneself and one's ability.
Organization: Maintains order and plans out work.
Time Management: Works efficiently.
Independence: Works effectively without supervision
Critical Thinking: Conceptualizes, analyzes, synthesizes, evaluates, and applies information.
Analytical Thinking: Breaks down complex problems into parts to arrive at a solution.
Reflective Thinking: Solves problems by examining of past experiences.
Creativity: Combines previously uncombined elements to create new knowledge.
Judgment: Assesses situations and draws sound conclusions.
Decision Making: Effectively chooses between two or more paths of action.
Technical Ability: Is proficient in practical skills.

Interpersonal Skills
Deportment: Treats others with respect.
Involvement: Active participation in the program.
Written Communication: Expresses facts and ideas in a clear, convincing, organized manner.
Presentation Skills: Effectively provides information to a variety of audiences.
Verbal Communication: Facilitates an open exchange of ideas and dialogue.
Listening: Understands, clarifies and evaluates verbal communication.
Teamwork: Facilitates cooperation within groups.
Interdependence: Relies on others for information and resources.
Conflict Management: Is able to negotiate around and resolve other people's grievances.
Networking: Makes and uses contacts to build and strengthen opportunities.
Leadership: Influences the performance of others toward goal accomplishments.

Figure 2b: MBT 360 Evaluation Form

tion aims to provide a practical real-world biotechnology experience and the students are assessed on their ability to integrate all their learning acquired throughout the program.

At the beginning of the program, each student is assigned two genes of interest to study which form the basis of many assignments and examinations. Students are required to conduct a literature review and a research plan for a diagnostic and vaccine. A feasibility study is performed on a therapeutic gene of interest to make final recommendations regarding its usefulness as a therapeutic or diagnostic target. For the final exam, students are expected to develop their therapeutic or diagnostic concept from its novel stages through to pre-clinical trials.

Based on the results and feedback from the feasibility study report, students construct scientific posters outlining the development of their concepts and are assessed on the contents of the poster as well as their presentation. Similar to scientific conferences, the scientific poster examination is open to the public and gives students a platform to discuss what they have developed while allowing feedback from public and industry participants.

Concurrently with the scientific process, students are integrating their knowledge and skills from the business aspects course by writing a business plan about their developed concept. The business plan is presented to potential investors as another part of the final examination. In a board room setting, students are given five minutes in which to entice the potential investors to invest in their concept and company. They are assessed on their business plan as well as the business presentation.

Throughout the final examination process, students are assessed on their scientific and business knowledge, their understanding of the concepts, as well as their presentation skills, communication, organization, and creative and analytical thinking. Combining the core courses with aspects of the careers course ideally provides the students the opportunity to engage in biotechnology situations while employing everything they have acquired throughout the program.

THE PRACTICUM

The MBT *Practicum in Biomedical Technology* is the capstone experience for students in the program. Most practica are in industry settings. The practicum project has practical applications for the company involved as they would have expected to have to do this activity at some point. Examples of practica include feasibility studies, market analysis, business plan writing, patent writing, and project management. Laboratory-based practica must have some aspect of commercial applications of the work, such as target validation, diagnostics development, optimization and validation. In recent years, clinical research and trial applications and cost analysis of clinical trials have been of particular interest to some MBT students.

Whatever the project is, there has to be an element of critical analysis related to biomedical technology and this element must be defensible. The project deliverables to the hosting company or lab may therefore be quite different from the student's practicum report. There is a defence of the practicum report following the experience. In this exam, much like a candidacy exam, students are expected to present their practicum experience, discuss their deliverables, and critique their role and the issues around the project they did. Approximately one quarter of practicum placements become permanent positions for the graduate or lead them through the company's network to another position in a similar role.

MBT ALUMNI

The MBT program makes an effort of staying connected with alumni to track their progress and professional endeavours. As of November 2007, 100 students have graduated from the MBT program since its creation in 2000 and the current whereabouts of 90 percent are known. A recent MBT alumni survey showed the majority of recent graduates are currently involved in biotechnology related businesses, suggesting that the program is meeting its objectives of preparing students for biotechnology careers. Earlier graduates had a harder time understanding or explaining what the

MBT meant, yet many of them have gone on to become leaders in biotech. Of these, three have started their own companies and several others are consultants or working in commercialization offices.

One of the greatest measures of success of MBT graduates is the number who have gone on to receive post-grad business internships (10 of 100) to continue working in a company setting. Ninety percent of alumni respondents in our recent survey would recommend the MBT program to a friend, specifically those interested in biotechnology or in learning the fundamental theories and concepts of biotechnology.

At the conclusion of each year, MBT faculty members and students meet to evaluate the year. Exit surveys are distributed to graduating students to receive feedback on aspects of the program including course content, program resources, overall quality of the program, and post-graduation plans. Furthermore a SWOT (strengths, weaknesses, opportunities, and threats) analysis of the program is performed by faculty members along with student representatives. These analyses, exit surveys, and alumni surveys are incorporated into one comprehensive final evaluation. The document and the recommendations therein are implemented whenever possible. It was feedback from faculty, alumni and then current students that led to the creation of the core science courses in 2002 and of the business aspects and career skills courses in 2004. The MBT program feels that this process is essential to ensure that program goals and objectives are being met, as well as helping to identify issues in the program that need improvement.

CHALLENGES

The strong oil and gas industry and its prominent role in Alberta's economy presents a challenge to biotechnology companies and educational programs, such as the MBT program. They are often overlooked by investors, and have difficulty gaining recognition and support within the province. The MBT program has managed to develop strong, sustainable, connections with industry, which are bringing recognition to the program and to the Alberta biotechnology industry as a whole. This will ultimately lead to a more sustainable and successful future for biotechnology in Alberta. Although the MBT program is a relatively new program, its recognition continues to grow due in part to its industry participants, networking events and largely due to the success of its graduates.

FUTURE DIRECTIONS

Keeping a small enrolment will remain an important aspect of the program. The student enrolment is capped each year to ensure an intimate learning experience and environment for its students. The program is, however, looking to expand internationally by making connections with international biotechnology industries, institutes and universities from countries including the United States, India, Brazil, Australia, and Chile. Creating international affiliations and agreements will not only improve the contacts and connections of the program, but will provide an opportunity for its faculty and students to understand the global impact of biotechnology. An ultimate goal for the program would be to seek international recognition on par with an MBA.

Over the past two years, the faculty and its associates have been working on a proposal to build an MBA/PhD program to be supported by CREATE (Centre for Research, Entrepreneurship, and Applied Technology Education). CREATE arose in response to the desire of the Government of Alberta to develop Centers of Excellence in research entrepreneurship and commercialization. The proposed program will integrate the business aspects similar to the MBT curriculum with the science focus of a master's or PhD level research degree, while at the same time addressing the existing and emerging economic drivers of the Alberta economy. Support from the University of Calgary, industry and government has been strong and the program recently received funding which will allow it to commence in 2009. CREATE provides both an exciting and challenging task for the faculty involved and will bring the efforts and vision of the MBT program to the next level.

CONCLUSION

From its creation in 2000 through its evolution to the program it is today, the MBT program is proud of its students and their success in the biotechnology industry. The evolving nature of the program allows it to remain current with the issues in the biotechnology industry while taking into account the needs and expectations of its students. It is the feedback from its students, faculty, alumni and local industry that has made the MBT program what it is today and will shape what it will become in the future.

Creating 'Business-savvy' Bioscientists at the University of Auckland

Joerg Kistler, Wendell Dunn, Claire McGowan, and Geoff Whitcher

Joerg Kistler, Ph.D., is Professor of Cellular & Molecular Biology and Director of the School of Biological Sciences at the University of Auckland. Joerg can be contacted at *j.kistler@auckland.ac.nz.*

Wendell Dunn, Ph.D., is Foundation Professor of Entrepreneurship and Associate Dean for Academic Development at the University of Auckland Business School. Wendell can be contacted at *w.dunn@auckland.ac.nz.*

Claire McGowan, Ph.D., MBA, is Director of the Bioscience Enterprise Programme at the University of Auckland. Claire can be contacted at *director@bioscienterprise.ac.nz.*

Geoff Whitcher is Commercial Director of University of Auckland Developments. Geoff can be contacted at *g.whitcher@auckland.ac.nz.*

Business and science often seem two different worlds. As subjects they are generally taught in separate programs and faculties. Yet employers in both the public and private sectors increasingly seek graduates skilled in both science and business.

The world economy is undergoing significant change, with increasing emphasis on the creation and application of knowledge as the foundation of economic prosperity. Discovery, creativity, and innovation are necessary but not sufficient conditions for wealth creation and economic transformation. The latter requires people having an understanding of both commercialization processes and entrepreneurship, including appropriate risk-taking.

Contributing to the demand for business-savvy scientists is the volume of research and development currently being undertaken by smaller technology companies and start-ups. Greater numbers of professionals are required who understand the entire value chain from discovery to market. This is particularly true in biotechnology, which is experiencing strong growth globally, and where innovation now occurs more often at universities and smaller research companies than, as was traditionally the norm, in large international corporations. The Bioscience Enterprise post-graduate program at the University of Auckland (New Zealand) addresses this demand by 1) adding business and legal skills to the armamentarium of new life sciences graduates, and 2) retrofitting those same skills into young but experienced scientists having industry experience. The program was launched in early 2006 and is taught in partnership between the School of Biological Sciences, the Business School, and the Law School.

The program consists of two one-year modules, with the first year carrying it own qualification and leading into a sequential but separate qualification. Students completing the first year successfully receive a Postgraduate Diploma in Bioscience Enterprise (PGDipBioEnt). Those achieving an appropriate standard may enter the second year leading to a Master of Bioscience Enterprise degree (MBioEnt). The entry requirement for the PGDipBioEnt is a bachelor's degree in any area of the life sciences or related disciplines. Entry is limited to around 20 full-time and up to ten part-time students and all are selected based on both academic record and an informal interview. Based on current experience, twelve to fifteen students from the diploma cohort continue into the master's program each year.

The PGDipBioEnt consists of five mandatory core and three elective taught courses. The core covers accounting and finance, marketing for scientific and technical personnel, emerging applications in biotechnology, intellectual property and relevant legal frameworks, and the commercialization of research. Courses, taught by a combination of academics and industry practitioners, employ a variety of pedagogies including lectures, case studies, and project work. Electives may be chosen from graduate programs in any life science-related area taught at the University of Auckland. At the end of their first year, students desiring to enter the master's program are strongly encouraged to undertake a summer internship in the regional biotechnology industry.

The master's year consists of three parts:
1. Short, intensive modular courses on product development and regulatory environments, and contemporary developments impacting the biotechnology industries
2. A research methodology module
3. A formal research-based master's thesis. Following completion of the first two parts, students enter a six-month internship with a company or organization to carry out a business research project, supervised jointly by one academic and one industry expert, the core issues of which form the basis for the thesis

Students are required to report formally on their progress at each of three milestone meetings and to submit their thesis for both internal examination and external assessment.

Auckland's Bioscience Enterprise program is accompanied by bi-weekly forums, each of which usually consists of a seminar given by an industry executive, followed by a networking function. The forums are widely advertised and designed to be educational and inspirational and to provide networking opportunities. A mixed audience of people attend from the biotech business community (e.g., entrepreneurs, biotechnology and other relevant firm principals, investors, professional service providers) in addition to university staff and students and visitors from other research institutions.

The program is well-embedded as a part of the University of Auckland's greater entrepreneurial ecosystem[1] to ensure it remains market-informed and that industry gets the multi-skilled work-

---

1   *Spicer, B, Dunn, W and Whitcher, G, "Transforming knowledge into wealth in a New Zealand research university" in "Toward an Ecosystem for Innovation – Implications for Management, Policy, and Higher Education", Special Issue, Industry and Higher Education (IHE), August 2006: 243-248*

force it critically needs for continued innovation and growth. As a result, long-standing boundaries between industry and academia are disappearing as increased interactions increase the knowledge of, and the understanding between, parties, encourage public-private collaboration, and bilaterally foster a spirit of dynamism and energy.

NEW ZEALAND BIOTECHNOLOGY AND PROGRAM CONTEXT

Following its re-election in 2002, New Zealand's Labour-led government launched a broad national economic development strategy based in part upon a Growth & Innovation Framework emphasizing three priority areas: information and communications technology, biotechnology, and creative industries such as design and filmmaking. In May 2003, delivery of a Biotechnology Taskforce Report on the future and potential of the field led to the creation of a *New Zealand Biotechnology Strategy*.[2,3]

The Strategy addressed a number of objectives—e.g., strengthening capability in science education, basic research, enterprise and commercial skills, skill-building around areas of strength, and promoting biotechnology for economic transformation and sustainability—and set in place a goal of increasing the number of core biotech companies from around 40 at the time to some 200 by year 2013, thus placing considerable emphasis on (new commercial) skills development. This created not only a high-priority need for such skills and experience, but a catalyst and rationale for new program development:

New Zealand needs to have 'commercially savvy' scientists and 'scientifically savvy' industry managers. This calls for implementing cross-disciplinary links at tertiary institutions.

2   *"Growing the biotechnology sector in New Zealand – a framework for action", Report from the Biotechnology Taskforce, NZ Ministry of Economic Development, May 2003*

3   *New Zealand Biotechnology Strategy – a foundation for development with care, NZ Ministry of Economic Development, May 2003*

New Zealand's biotechnology industries have flourished in the period following formulation of the Strategy, despite a continuing lack of appropriate management skills, especially in marketing.[4] Indeed, the L.E.K. *New Zealand Biotechnology Industry Growth Report* painted a picture of robust growth and diversity in the sector.[5]

In 2006, for example, the annual expenditure on biotechnology activities in New Zealand was approximately NZ$640mm, of which about NZ$200mm was funded by government. There are now more than 2,200 people employed in biotech-related jobs in roughly 125 private and public sector entities, including institutions of higher education. Local venture capital funds are growing rapidly, the largest currently having attracted some NZ$150mm to invest in the sector.

New Zealand biotechnology activities cover a range of applications. About 40 percent of the 2006 total biotechnology sector activity (measured in terms of average expenditures and/or employment) is concentrated in agricultural biotechnology, adding value to what remains, predominantly, a commodity-driven export sector. Biotechnology is estimated to contribute around NZ$300-400mm per year to the total national primary sector earnings across its pastoral, forestry, and horticulture sub-sectors. Other NZ biotech sectors, while economically smaller in comparison to agriculture—medical devices/diagnostics (23 percent of biotechnology total), industrial applications (20 percent), and human health (17 percent)—are rapidly becoming significant in their own right.

While biotechnology activities occur across New Zealand, a major drug discovery and development cluster has formed around its largest city, Auckland, which has about 1.4mm of New

---

4   *Organisation for Economic Co-Operation and Development (2007) 'OECD Reviews of Innovation Policy: New Zealand', ISBN 978-92-64-03760-1 , BrowseIt edition downloaded as http://213.253.134.43/ oecd/pdfs/browseit/9207071E.PDF on 3 January 2008.*

5   *L.E.K. Consulting (2006) The New Zealand Biotechnology Industry Growth Report 2006, NZ Trade and Enterprise*

Zealand's 4.1mm inhabitants. Greater Auckland is clearly the national hub for pharmaceutical biotechnology, based in large part on the world-class biomedical research undertaken at the University of Auckland.

The University of Auckland is the country's premier tertiary educational institution, has roughly 38,000 students, and was ranked 50th among world universities in the 2007 *Times Higher Education Supplement* rankings; it stands 26th in the 2006 *THES* ranking of biomedical universities. Biomedical research, a major focus of the university, is concentrated in its School of Biological Sciences (Faculty of Science) and School of Medical Sciences (Faculty of Medical & Health Sciences). Its five major research centers and institutes have a strong research focus on drug discovery and medical devices.

Research outcomes have been highly productive and most of the leading biotech drug discovery and development companies in the region are, directly or indirectly, spin-outs from the university. These include publicly listed companies Neuren, Genesis R&D, and Living Cell Technologies, as well as private equity-funded companies Protemix, Proacta, and CODA Therapeutics. Three drug candidates are currently in Phase III clinical trials; eight are in Phase II; six are in Phase I. In addition, Auckland is the location of Fisher & Paykel Healthcare, NZ's largest medical devices innovator-manufacturer, with a total market capitalization of roughly NZ$2.4 billion.

Many of these companies are involved in the Bioscience Enterprise program. The relationships afford their professionals opportunities to contribute to program lectures and tutorials, as well as internship opportunities for the program's second-year students. Past activity surrounding these companies—and which is anticipated for several new ones about to be established—set the scene for the development and initiation of the program in Auckland.

GRADUATE AND EMPLOYMENT PROFILES

Employment for program graduates is intended primarily to be entry-level roles in bioscience enterprises. Typically, these include 1) positions in marketing, market-analysis, product development, regulatory affairs, and business development in both commercial and non-commercial entities (e.g., biotech ventures, pharmaceutical, medical device, diagnostic and reagent manufacturers, the food and beverages industry, technology transfer and licensing offices, and government research organizations), and 2) technical analyst positions in business development firms, finance, investment and venture capital firms, and in government agencies and ministries.

Given this very broad employment profile the objectives in designing the MBioEnt program were to educate and train graduates who 1) had a solid grasp of the fundamental principles of contemporary research in the life sciences, 2) were capable of recognizing the potential of high value products or processes developed from basic discoveries, and 3) understood how to create, protect, and manage the underlying intellectual property. Graduates were to be able to work in interdisciplinary teams to support the development of strategies for commercializing research and to secure the capital resources required. They would need to understand the needs of a variety of organizations—for profit and not-for-profit—in performing research valuations and carrying out due diligence studies, whether for start-up or advanced companies, communicate and present well, and be able to move confidently in both the science and business worlds.

PROGRAM DEVELOPMENT AND FUNDING

As a consequence of the May 2003 *New Zealand Biotechnology Strategy*, New Zealand Trade and Enterprise (NZTE) commenced sponsorship of a stream of high-profile biotechnology entrepreneurs from the USA and Europe. Its intent was to raise awareness of, and to engage in discussions about what could be done to grow, New Zealand's fledgling biotech sector.

One focal point was a critical lack of local capital with which to develop the industry. It was noted that this was due to a distinct lack of understanding of biotechnology by the local investment community and a lack of knowledge by local biotechnologists of commercialization processes and investor expectations. There was no common platform of understanding between parties to build the knowledge and trust required to complete financing and enable company development.

At that same time, the University of Auckland business school's newly arrived inaugural Professor of Entrepreneurship, together with the head of the School of Biological Sciences (a member of the chair search committee), ran a summer course on biotechnology research commercialization, the delivery of which promised to test institutional flexibility. A for-credit graduate course, *BIOSCI 745: Biotechnology and Bioentrepreneurship*, was assembled with considerable haste and delivered in the 2004 antipodean summer.

The course was designed for two different audiences—academic and industrial—and consisted of six sessions delivered over three weeks in two parts serving two interest streams. Each session consisted of a public lecture plus a follow-on set of discussions, tutorials, and practical exercises. Limits were set: 50 students who could enroll formally for credit, and another 50 industry and related professional community participants who could attend the public lecture as guests of The University of Auckland.[6] About half of the fee-paying enrollees were graduate students; the other half from the university and industry. Two streams were offered: one for the scientists who were taught the basics of business, and one for business people who learned some basics of biotechnology. Public lectures were presented by visiting high-level US biotechnology entrepreneurs and experts, funded by NZTE. A team of local industry practitioners and academics delivered many of the for-credit tutorials and practical exercises.

The course was a huge, resounding, success. Not only was the credit portion fully subscribed, the free evening public lecturers

---

6   *Included among the students enrolling in (and passing) the formal credit course were the head of school and the chancellor of the university.*

were far over-subscribed, with some attracting audiences well beyond advertised room capacity. The course achieved a very high profile throughout local industry and legal circles, further enhancing the university's reputation for a developing entrepreneurial culture and rigorous, market-informed, education.

Following this summer school success and an assessment of what other types of programs might contribute the most value to NZ biotechnology industry, the now-expanded local delivery team committed to create a two-year master's program in Bioscience Enterprise.

The local biotechnology industry was consulted extensively in developing program content to ensure that the proposed graduates (see *graduate and employment profiles*, above) would meet market needs and expectations. Much was learned by visiting the teaching team of the same-named program at the University of Cambridge, UK, who shared generously their program information and experiences.

In late 2005, the Committee on University Academic Programs (CUAP) of the New Zealand Vice-Chancellors Committee approved the new Auckland program for the 2006 academic year; the PGDipBioEnt launched the following semester; March 2006. In March 2007 the Master's was launched following completion of the diploma program by its first cohort.

Development costs were borne in part by The University of Auckland and in part by the government. Because the program promised significant new human capital creation for the nation's rapidly emerging biotech industry, application was made to the Growth & Innovation Pilot Initiative (GIPI) fund of the Tertiary Education Commission. The application was successful and contributed approximately NZ$900K to support development and implementation of the first two years of the program. While delivery costs largely are covered through student tuition and fees plus government bulk funding, GIPI funding proved quite valuable in covering the innumerable and unpredictable costs of developing a cutting-edge program crossing both conventional disciplinary and administrative boundaries.

The development process itself proved a most pleasant experience. The original course development team was expanded to twelve members by including academics representing all the relevant disciplines from the School of Biological Sciences, the Business School, and the Law School, as well as an NZTE representative and the Director of NZBio/Auckland industry group.[7] The team met monthly, developing true camaraderie across long-established faculty boundaries. Near the conclusion of the development process, all three participating faculties signed a memorandum of understanding to 'own' and deliver the program jointly.

Following the initial program development phase, a program director was employed; the initial three years was funded from the GIPI grant. The position—Director, Bioscience Enterprise Programme—is, by intent, a half-time appointment, with the individual holding a concurrent executive position in the biotech industry. This dual-by-design role has proven vitally important in sourcing industry lecturers, guest speakers, and student internships, as well as in maintaining overall voice-of-market reality and connectedness.

The new programs were marketed through parallel promotional mechanisms: news articles in the local media, a color brochure—widely circulated through the university community and the biotech industry, accompanied by stories or sidebars in various "house organs"—and, digitally, using internationally oriented web pages. On its 2006 launch, the program received a mention in a feature on entrepreneurship programs in *Nature*.[8] Following its promotion, program pages received regular hits from offshore inquirers, confirming relatively easy, if not direct information access by Google-ing the Faculty of Science web pages on the University website.[9]

---

7   *The Business School's academic staff in particular—most veterans of high profile international MBA programs—brought and shared valuable experience.*

8   *Rentzsch, Robert (2006) Special report: After-school programs Nature 440:122-123*

9   *www.auckland.ac.nz*

The official Bioscience Enterprise programme web page is: *http://www.science.auckland.ac.nz/uoa/science/about/subjects/bioscience_enterprise.cfm.*

ACADEMIC-INDUSTRY LINKAGES

The Bioscience Enterprise program relies heavily on close linkages with the biotechnology industry and, in turn, provides many opportunities for strengthening such ties. Not surprisingly, a significant proportion of the program's initial uptake and success can be attributed to the social capital and personal reputations of its developers. These include not only the high-visibility academics involved, but also others in the development team such as the program director and the commercial director of University of Auckland Developments, an early member of the program development team.

For example, the commercial director and another member of the program development team were founding members of the regional biotechnology industry body, NZBio/*Auckland*, and both continue their active presence. Similarly, the current program director was previously director of NZBio/*Auckland*. Collectively, their work facilitates industry interactions, helps ensure that the university enjoys a very high profile among NZBio members and industry entities and plays—and is seen as playing—its role in national biotechnology knowledge-workforce development.

Just as the program directors' personal reputations and contacts within the industry have played fundamental, if not pivotal, roles in the program, so too have various other activities and events directly associated with the University of Auckland's still-evolving entrepreneurial culture. Many were developed, led, or supported by the commercial director in concert with the head of school, the program director, the entrepreneurship professor, and the business school dean.

Examples include:
- Bioscience Enterprise Forum: Twelve early evening seminar-events presented at the School of Biological

Sciences by an industry executive and followed by a networking social function. Sponsored by New Zealand Trade & Enterprise, each forum is advertised to industry through the NZBio newsletter, thus attracting a steady stream of people from the business community into the university. Attendance by all program students is mandatory as each event offers excellent networking opportunities.

- Chiasma[10]: a student-led network within the University of Auckland was formed to promote closer links between industry and the student biotechnology and life sciences communities. Chiasma, which is well-connected with the Bioscience Enterprise programs, organizes regular bioentrepreneurship competitions and career evenings, all drawing heavily on industry participation. The network is supported in part through the GIPI grant.

- SPARK[11]: the University-wide entrepreneurship challenge competition, was established in 2003 to foster a spirit of innovation and enterprise across the institution and to encourage the establishment of new ventures based on university-generated research and ideas. Judging panels consist generally of industry practitioners and the competitions, which enjoy wide industry financial support, regularly feature entries from biotech and related science students.

- The ICEHOUSE[12]: the business incubator of the University of Auckland and provides business development support for start-ups which may be co-located in the university's research laboratories

---

10 www.chiasma.auckland.ac.nz
11 www.spark.auckland.ac.nz
12 www.theicehouse.co.nz

where appropriate. Representatives from both
UniServices and the ICEHOUSE regularly
contribute to the Spark activities and the Bioscience
Enterprise program.

These, in turn, are complemented by major University of
Auckland-based entities and initiatives, such as the following,
which generate and support biotech-related innovation:

- The Institute for Innovation in Biotechnology[13],
  located in the School of Biological Sciences,
  was established in 2006 to facilitate academic-
  industry partnerships and provide working space
  for innovative biotechnology start-ups and co-
  located companies. IIB capital funding (as proposed
  to government by the original developers of the
  bioscience enterprise program) ultimately will total
  NZ$32 million.

- Auckland UniServices, Ltd.[14] The University
  of Auckland has a strong track record of
  commercializing its research, particularly in
  biotechnology and pharmaceuticals for which it
  has well-established international connections.
  UniServices, its commercial arm, is the largest
  contract research and technology transfer office
  in Australasia, having annual revenues in excess
  of NZ$70 million. Several biotech start-ups and
  mature companies, co-located with the School
  of Biological Sciences and the School of Medical
  Sciences, offer students jointly supervised post-
  graduate projects as well as opportunities for
  employment.

Taken as a whole, the emerging picture is one of inter-depen-
dency; an entrepreneurial ecosystem in which traditional bound-

13   *www.biotech.co.nz*
14   *www.uniservices.co.nz*

aries between industry and academia are blurring in continuum. The Bioscience Enterprise program sits amidst a shared community of interest, ensuring both that the program remains market-informed, and that New Zealand's biotech industry can access the workforce skills it requires for continuing growth and competitiveness.

AUCKLAND'S BIOSCIENCE ENTERPRISE PROGRAMS IN DETAIL

1. POST-GRADUATE DIPLOMA IN BIOSCIENCE ENTERPRISE (PGDIPBIOENT)

Candidates for entry into the PGDipBioEnt must have a bachelors degree related to the life sciences (e.g. bachelor of science with a major or specialization in biological sciences, bioinformatics, biomedical science, food science, medicinal chemistry, pharmacology, or psychology; a bachelor of biomedical engineering; a bachelor of pharmacy; or, a bachelor of biotechnology.)

Enrolment may be full-time (first semester start only) or part-time. The latter appeals to employees who wish to upskill and extend their knowledge base. In each of its two operational years (2006, 2007) the PGDipBioEnt class had 20 full-time and seven part-time students. Most full-time students are pre-experience—i.e., progressing directly from their undergraduate program without work experience. Employed part-time students add significant value as they can share their industry experiences with their classmates and add a sense of commercial reality.

Enrolment into the PGDipBioEnt is competitive; the program is limited to a maximum of 30 full-time and part-time students each year. Applicants are ranked based on academic achievements, but relevant industry experience is taken into account, and each applicant undergoes an informal interview to assess their motivation for the program.

PGDipBioEnt studies require one year full-time-equivalent study: 120 total course points. Students enroll in five compulsory core and three elective courses; each of 15 points. Electives must be graduate level and may be chosen from a variety of programs across the full life sciences spectrum. The advice to students is that they select courses which build on their existing knowledge foundations and engage their professional interest.

The program's academic year consists of two semesters of twelve teaching weeks each. Each course generally is assumed to require at least ten hours work per week (120 hours); 40 hours-plus for a four-course, full-time, load (A general 10X 'rule of thumb' suggests that a fifteen-point course require a minimum 150 hours in toto for class, projects, study, and examinations). Most courses have a once-weekly class session of three hours for formal lectures, case study discussions, guest lectures and/or student presentations. The latter are based on assignments, individually or in groups, which require analysis or project work.

Compulsory core courses are held on weekday evenings, generally from 16:00 to 19:00. Such scheduling: 1) minimizes clashes with day-based electives, 2) makes it easier for students in employment to attend, 3) facilitates recruitment and availability of practitioners for lectures and tutorials, and 4) acclimatizes students to working hours beyond the 'normal'; a situation they doubtlessly will encounter regularly during their professional lives.

The five compulsory courses listed for the 2008 PGDipBioEnt program core are:

SCIENT 701 Accounting and Finance for Scientists (semester one)
Builds on scientific numeracy in exploring the sources, uses, and reporting of accounting and financial information in science-based enterprises; application of capital budgeting and valuation theory to science-relevant situations; and, key bases for financially informed project and enterprise decision-making and the management of economic resources.

SCIENT 702 Marketing for Scientific and Technical Personnel (semester one)
Examines the intermediaries and end-users of technical and research-related applications, products and services; their customers, value chain, and related concepts in both highly-regulated and

open markets; and, how effective science-related marketing strategies and promotional efforts are developed and communicated.

SCIENT 703 Frontiers of Biotechnology (semester two)

An examination of how breakthrough discoveries in contemporary life sciences flow through to commercialization; current and emerging applications of biotechnology; includes guest lectures from New Zealand's leading biotechnologists and case studies focused particularly on medical applications.

SCIENT 704 Law and Intellectual Property (semester two)

An explanation of the legal system including basic concepts of contract and corporate law in a biotechnology context. Emphasis will be upon intellectual property laws in particular patent law and practice and other means of protecting new ideas, discoveries and inventions. Also covered will be technology licensing and basic competition and marketing law.

SCIENT 705 Research Commercialisation (semester two)

Integrative exploration of common theories, processes and models involved in commercializing scientific research. Topics include technology transfer, technological entrepreneurship, commercial potential, risk, and valuation assessment and related tools. Utilises multiple learning approaches including case studies and a hands-on term project.

2. MASTER OF BIOSCIENCE ENTERPRISE (MBIOENT)

Entry into the master's program requires completion of the PGDipBioEnt with a minimum B average in at least 90 points (best six courses of the program). Of the sixteen inaugural PGDipBioEnt class of 2006 graduates who achieved this standard; fourteen enrolled immediately in the Master's program. Final results for the 2007 entering cohort are not yet known, but the progression rate is expected to be similar.

The MBioEnt program has a point value of 120 and requires one additional year of study. In addition to commencing their thesis development in early semester one, students take two compulsory courses, each of fifteen points and in modular format:

SCIENT 721 Product Development and Regulatory Environments
Aims to give students an understanding of the stages of product development for therapeutics, diagnostics and medical devices, as well as regulatory requirements affecting product development in the life sciences. Project management tools and processes will also be covered in the context of product development.

SCIENT 722 Current Issues in Bioscience Enterprise
An exploration of trends and development of importance to life sciences-related enterprises and industries. Utilises multiple learning approaches—e.g. independent readings, case studies, projects, guest speakers, presentations, and related discussions.

The main component of the MBioEnt is a formal master's thesis of 90 points (75 percent FTE). While *SCIENT 722* is intended both to broaden all students horizons and to help narrow their focal areas of research interest, appropriate formal research methodology and project management studies also are embedded in each candidate's thesis development program.

The thesis itself is based on a science-business research project carried out during a six-month internship in a company or government organization. Students submit a brief biographical sketch and declare their preferences for the internship from a portfolio of participating companies/organizations. Placements are finalized following a round of interviews. The research topic is developed between the student, an academic supervisor, and an industry supervisor. Students submit a project outline to the program director after the first month and present progress seminars at two milestone meetings which bring all interns and supervisors together. Following thesis submission, students present their projects at a final meeting. All meetings are held under confidentiality and, if necessary, theses may be embargoed for up to two years to protect intellectual property or commercially sensitive information.

LESSONS LEARNED – IMPROVEMENTS MADE

The Bioscience Enterprise program at the University of Auckland is the first of its kind in New Zealand (and, most probably, in Australasia) making it impossible to compare experiences nationally or regionally. Similarly, as most extant graduate award business-biotechnology programs have a wholly different structure and approach—most are sub-parts or tracks within conventional MBA programs, particularly in the USA—direct comparisons also are difficult. However, many valuable insights were gained from the same-named program at the University of Cambridge UK which commenced in October 2002.[15] While that program was not used as a *per se* model in developing the Auckland program, its leadership and lecturing team generously shared their own lessons learned.

Continuing program development and improvements at the University of Auckland have relied to a large extent on feedback from the students and their academic and industry supervisors, the lecturing team, visiting scholars, and web-based reconnaissance.

The principal lessons learned and improvements made during the program's first two years include:

- Acculturation: The learning step science students take in adapting to business-oriented thinking and perspectives involves a huge culture change. As most students enrolling in the program were unsure what was to be expected of them, a compulsory two-day introductory workshop was introduced.
- Pedagogy: Students appear to learn business thinking most effectively through management (decision-based) and historical (law) case studies and group fieldwork projects rather than conventional lectures. Some rebalancing of theory and practice was required in a few courses.
- Business-based internships: In developing

---

15 *The Cambridge program is of nine months duration and employs a 30-point project rather than a thesis.*

the program, an assumption was made that approximately half of the internships would result in a business-intensive thesis; the remainder a technical thesis. Generally, firms sponsored internships with commercial motivations. As a result, roughly 90 percent of students ultimately pursued a business-based thesis. Swift recruitment of a business faculty member to oversee the thesis process, along with several staff who—luckily—were actively interested in biotechnology-related or international research, resolved the skill shortage. Staff workload planning should address the issue in future.

- Business research methodology: Master's students were generally ill-prepared for business-focused research, especially business methods. Attendance at a two-day introductory workshop at the start of the master's year plus participation in an applied research methods course (not for credit, being a part of the thesis) will become mandatory from 2008. The combination, plus changes in the *SCIENT 722* course structure, should provide students adequate additional research tools and frameworks.

- Rigor-relevance balance: Managing the professional and commercial expectations of industry sponsor-supervisors, given the substantial academic depth demanded of graduate students by university supervisors (and NZ educational standards), has required good and frequent communications amongst all parties. It is essential to set expectations early and trilaterally in defining student industry internships.

- Administrative leadership: The program director plays a central, if not pivotal, role in the success of the program. A high-level hire is absolutely essential, and the best type person for the role appears to be a dual qualified professional (science and business) with executive and/or consulting experience and extensive contacts throughout the industry. As this director manages virtually all aspects of the academic-industry interface (e.g., the forums, industry speakers and events, and internship placements) and is the principal point of contact for all students in the program—particularly when they are dispersed during their (potentially international) internships—it is essential that the individual be personally and professionally suited for this most demanding role, and rewarded accordingly. Turnover in this role, given that the individual has concurrent industry responsibilities, has a very real potential to place the program and its widely-dispersed thesis-stage students at risk.

- Networking: Regular encounters with industry contacts have proven essential in placing students for internships, masters research, and employment as well as supplementing their practical education. Such networking has afforded excellent opportunities and linkages, and raised industry awareness of a 'new breed' entering their employment market; graduates fluent in business and science. To supplement the program forum events, all master's students were provided free admission to the annual NZ Biotech Conference as well as free membership in NZBio. Such student support will continue as a permanent feature of the program.

MEASURING SUCCESS

At the time of writing, many in the first cohort of master's students are still in their internships, working on their thesis projects. Although several have submitted their finished theses, we lack concrete data on the types of jobs program graduates ultimately will find.

The fact that internship placements were readily found for all fourteen students in the first MBioEnt cohort is clearly a positive outcome. Organizations offering internships included a large medical devices company (taking two students), a multinational pharmaceutical firm, a technical university, several local biotechnology start-ups, two Crown Research Institutes, a merchant bank, as well as NZ Trade & Enterprise and the Ministry of Research, Science & Technology.

The industry already is making enquiries about students availability for summer internships during the 2007/2008 Christmas break. They are also inquiring about student availability for thesis-based research internships (May start) in the 2008 academic year. Internship take-up (placements, speed, inquiries) appears a valid and useful metric.

Industry feedback on many of the interns has been very positive. Several students already have received, before thesis completion, job offers from their internship companies. Two non-progressing PGDipBioEnt graduates have found employment in a science-business area and one part-time student has been promoted by her employer into a new position in their IP team. Law firms are watching intently to assess the potential for further training of graduates as patent agents or attorneys. The program's alumni database is growing, and the intention to maintain contact with alumni and employers, and to monitor career outcomes, is well-formed.

Student feedback from courses and field internships is both a valuable and fair measure. Course evaluations have returned satisfaction ratings ranging between 80 and 100 percent for all courses save one. The latter, *Research Commercialisation*, was criticized for having too much theory at the expense of practice. The course has been totally redeveloped using case studies and, in its new format, is receiving excellent student feedback at the time of writing.

Finally, the willingness of industry professionals to give of their valuable time to lecture in the bioscience academic program or to speak at the bioscience forums can be taken as further measures of program success, as can the number of industry people who attend the forums. These factors all indicate strong and continuing industry interest in, and support for, the program and are fair measures of program success to date. Such close ties between academia and industry should help ensure that the Bioscience Enterprise program at The University of Auckland continues to be market-informed and on the cutting edge.

Masters of Science in Biotechnology Entrepreneurship Program: One of the Professional Master's in Science and Technology Entrepreneurship Programs at Case Western Reserve University

Christopher A Cullis

Christopher A Cullis, Ph.D., is Francis Hobart Herrick Professor of Biology and Director of the MS in Biotechnology Entrepreneurship program at Case Western Reserve University. Christopher can be contacted at *cac5@case.edu*.

A focus on biotechnology (and associated health care and medical devices) is a major area of interest to northeast Ohio. The wealth of activity within the region including Case Western Reserve University (CWRU), the Cleveland Clinic, and university hospitals provides opportunities to grow a biotechnology cluster in the region. One method of increasing this activity is to improve the talent available in all areas important for the development of startup and fledgling companies, including scientific, managerial and marketing expertise. Funding in the form of the Third Frontier Program has provided additional resources for growth, with 43 percent of the funds in this program being targeted to the biosciences. The MS in biotechnology entrepreneurship was an opportunity to augment the expertise in this area as well as providing an additional route to the commercialization of research results from the large research enterprise in the biological sciences within the region.

UNIVERSITY CONSIDERATIONS

The professional master's in Science and Technology Entrepreneurship Programs (STEP) at CWRU were initiated with the physics entrepreneurship program. The success of this program led to an expansion of the offerings to include equivalent programs in biotechnology, chemistry, statistics and mathematics. The organization of all these masters programs included a four course core sequence, equally divided between courses in the discipline and in venture creation, as well as a one year internship. The master's thesis had to have sufficient scientific content to satisfy the departmental master's requirement as well as sufficient commercial analysis to provide a feasibility study of the project. One of the early hurdles was to convince the biology department faculty that these theses would contain sufficient science to make them eligible for the award of a master's degree in biology.

ORGANIZATION AND ADMINISTRATION OF THE PROGRAM

The program is administered within the biology department by the entrepreneurship program steering committee. This committee includes four biology department faculty—three of whom are, or have been, involved with biotechnology startup/spinout companies (Cullis, Caplan, and Haynesworth). Graduate applications are initially processed through the normal CWRU channels but the final admission and financial aid decisions are made by a separate biotechnology entrepreneurship graduate admissions committee. All of the rest of the degree requirements are processed as any other biology department master's degree. One necessary structural change was the requirement to add biotechnology entrepreneurship as an area of specialization for the master's examination. The introduction and continuation of this program has occurred without the addition of any new faculty to the department.

FUNDING OF THE PROGRAM

The initial funding of the professional master's in STEP was through a series of grants from individuals and organizations including:

- A bequest from Robert Stieglitz
- The Coleman Foundation
- The National Collegiate Inventors and Innovators Alliance
- The Alfred P. Sloan Foundation
- The Cleveland Foundation

Additional institutional support was in the form of tuition waivers during the first five years of the program. These waivers were on a reducing scale and acted as an incentive for students too enroll while the program was in its early stages. Any financial aid to students in the Biotechnology Entrepreneurship program is now competitive with all the other graduate programs within the biology department.

GOAL

The purpose of this program is to develop an understanding of modern molecular biology and biotechnology applied to the launching of new ventures in the biotechnology industry. In addition to advanced training in the fundamentals of biotechnology, courses will track the relationship between the academic innovation and its application, with an emphasis on the blurring of the distinction between academic and industrial research. The program will feature a 9-credit research thesis focused on the development of a business plan for a biotechnology initiative to introduce the students to the culture of small biotechnology startup companies.

OVERVIEW OF THE PROGRAM

The master's program in Entrepreneurial Biotechnology at CWRU connects students with the business executives, leaders, experts, and venture capitalists that are crucial to success in start-up and growing ventures. It is a 2-year degree offered by the Biology Department in conjunction with STEP. The program provides studies in state-of-the-art biotechnology and scientific innovation, practical business instruction, and real-world entrepreneurial experience to individuals with a bachelor's, master's, or PhD in a biology-related field.

The two-year program includes these main components:

- A core of courses taught by the Department of Biology including a two-semester sequence *Contemporary Biology and Biotechnology Innovation I* and *II*, specifically designed for this program.
- A core of courses taught in conjunction with the other STEP programs in new venture creation and technology entrepreneurship.
- A biology master's thesis involving an entrepreneurially-oriented project. This will typically arise from an entrepreneurially-oriented internship in a sponsor company, or from a student-designed research project that will be the basis for launching a new venture.
- Mentorship and activities that forge crucial relationships: Students will also be paired with a faculty supervisor/mentor who will guide them in the development and execution of their research projects, and who will be a long- term scientific resource. With the assistance of the STEP, the supervisor/mentor connects students with people who are crucial to the success of start-ups and growing ventures, such as business leaders, executives, intellectual property experts, successful entrepreneurs, and venture capitalists.

Highlights of the program include seminars, conferences, business plan competitions, extensive library databases for marketing, industry, business and science research, an entrepreneur-in-residence, and customizable internships through the Commercialization Assistant (CA) Program. In a short period of time, the student can develop a formidable network of connections.

CURRICULUM DETAILS

The curriculum of the MS Program in Biotechnology Entrepreneurship is a 30 credit-hour track, housed in the biology department. The curriculum includes four required courses (twelve hours), a nine credit hour thesis, and nine credit-hours of technical electives, three of which must be Biology course work.

The two required Biology courses are:
BIOL 491 - Contemporary Biology and Biotechnology for
        Innovation I (3 credit hours)
BIOL 492 - Contemporary Biology and Biotechnology for
        Innovation II (3 credit hours)

The Technical Elective is a 400-level or higher biology/biomedical sciences, biotechnology or informatics course or other technical elective appropriate to an individual student's program of study, as approved by the Biology Entrepreneurship Program committee.

The other two required courses in entrepreneurship are:
ENTP 429 - New Venture Creation (3 credit hours)
ENTP 441 - Technology Entrepreneurship (3 credit hours)

Both of these courses are taken in conjunction with all the other STEP students.

To complete the program, students will have option for 6 credit hours of course work in approved electives, which may be courses in science, engineering, management or accounting appropriate to an individual student's program of study, as approved by the Biology Entrepreneurship Program committee.

COURSES

Contemporary Biology and Biotechnology for Innovation I is the first half of a two-semester sequence that provides an understanding of biotechnology as a basis for successfully launching new high-technology ventures. The course will examine physical limitations to present technologies, and the use of biotechnology to identify potential opportunities for new venture creation. The course will provide experience in using biotechnology for both the identification of incremental improvements and the basis for alternative technologies. Case studies will be used to illustrate recent commercially successful (and unsuccessful) biotechnology-based venture creation, and will illustrate characteristics for success. The purpose of the course is to provide an understanding of contemporary biology and biotechnology as a basis for successfully launching new high-tech ventures.

Contemporary Biology and Biotechnology for Innovation II places more emphasis on current and prospective opportunities for Biotechnology Entrepreneurship. The purpose of the course is to explore the ways in which biology-based developments and inventions can impact society and thus provide an intellectual basis for successfully launching new high-technology ventures. The course will focus on the stages of biotechnology startup companies using models from the local area and having individuals from these companies provide the instruction. The topics will include the identification of novel technologies, the transfer of technology from the academic laboratory to a commercial concern and licensing and patent issues. Guest speakers for this course include principals from companies spun out of university laboratories, venture capitalists and technology transfer experts from CWRU.

The enrolment in this program is mainly full time students. However, in order to cater for a wider audience these courses are taught in the evening (6 – 8:30). Most frequently the majority of the students in these courses have not been students in the Biotechnology Entrepreneurship program. Both courses have enrolment from the Chemistry Entrepreneurship program.

Contemporary Biology and Biotechnology for Innovation I has enrolment from graduate students in the traditional biology master's and PhD programs, and graduate students from the biomedical engineering and computer science departments, among others. A small undergraduate enrolment also occurs. *Contemporary Biology and Biotechnology for Innovation II* was a required course for the MBA in Bioscience Entrepreneurship allowing synergistic interactions between science-based and management-based graduate students.

New Venture Creation focuses on the role of entrepreneurship in an economic unit that has been well-documented and is of interest to business people, politicians, and university professors and students. Creating and growing a new venture inside or outside the corporation is a task that few individuals are able to accomplish, even though many profess the desire. This course is based on an understanding of all the functional areas of business and applies the tools and analytical techniques of these functional areas to the new venture creation process in a domestic and international setting. The primary goal of this course is to provide an understanding of entrepreneurship and the entrepreneurial process in a global setting. The course will broaden a basic understanding obtained in the functional areas as they apply to new venture creation and growth.

Technology Entrepreneurship, which deals with the nature and importance of entrepreneurship in developing and transferring technology and positively impacting an economy, is an area of importance to business leaders, educators, politicians, and individual members of the society. To create something new and of value to both the organization and the market takes a technical individual who is willing to assume the social, psychic, and financial risks involved, and who achieves the resulting goals, whether they be monetary, personal satisfaction, or independence.

The objectives of these two courses are to enable the participants to understand all aspects of entrepreneurship and the entrepreneurial process, and to learn how this can be applied in a manufacturing organizational setting.

These two courses were originally taken during the first two semesters in the program. They are now being offered during the summer session and taken on entry to the program. This change in timing has allowed the students to earlier access to the commercialization assistant positions resulting in the possibility of completing the program in eighteen months rather than two years.

An additional boot camp to introduce accounting issues has also been an early communal activity in the Entrepreneurship Programs.

INTERNSHIPS

A feature of the program is the year-long internship, usually as a commercialization assistant associated with either a commercial entity or a university laboratory, which forms the basis of the master's thesis. This internship can be within an existing startup company or can involve the formation of a new company. An early successes was the formation of Superior Scientific Ltd. by participants in the first class of this program. Elsewhere, students have obtained internships in the new ventures in the biotechnology incubator run by BioEnterprise, a business formation, recruitment, and acceleration initiative designed to grow health care companies and commercialize bioscience technologies based in Cleveland. Students have frequently been employed by these companies following their graduation. In certain cases, students were already employees of some of these biotechnology companies and therefore the issue of internships was moot.

An additional sector for internships has been in university laboratories that have potential commercialization discoveries but the primary faculty members are not particularly interested in becoming involved in the commercialization process themselves. In these cases, the student projects have included a technology assessment in conjunction with the university technology transfer office and have subsequently been involved in the formation of new ventures. These internships give the students direct experience of the atmosphere and culture of biotechnology startup companies. They can also be supported in part from the Third Frontier Program in

Ohio, which will match a portion of the stipend of an intern in particular areas.

TECHNICAL ELECTIVES

These electives can be used by the students to strengthen particular areas of interests. Those students who enter the program with a weaker science background often take all of these electives as science courses. Some students, particularly those who are already in companies and have an extensive science background, take additional entrepreneurship or accounting electives to strengthen their management expertise. Thus, outside of the four required courses, the program allows a large degree of flexibility in student choices.

STUDENT RECRUITMENT

The students who apply for the program are a mix of both US and foreign students. Currently, the majority of applicants are foreign, with substantial numbers from India and China, and an increasing number from African countries.

OUTREACH

The director of the program has been involved in the development of a program at the University of Pretoria (see *K. Kunert*, this volume). The transfer of the experiences and the opportunities for exchanges of students between different settings in the USA and South Africa can provide exciting additional opportunities for the students in these programs.

FUTURE PLANS

The STEP and equivalent programs within CWRU include activities in the School of Engineering and the School of Medicine. The closer integration of these programs, especially with regard to the course offerings in venture creation and technology entrepreneurship will allow synergistic interactions across disciplines, especially in the areas of biomedical engineering and biotechnology.

Expansion of the biology program with formal participation of the School of Medicine will result in the admission of more students each year and the development of a larger number of new startup companies, resulting in additional opportunities for both internships and subsequent employment.

An Interdisciplinary Undergraduate Biotechnology Program at the University of Houston

Rupa Iyer

Rupa Iyer, Ph.D., is Director of Biotechnology Programs at the University of Houston. Rupa can be contacted at *riyer@uh.edu*.

In the last several decades we have made tremendous progress in recombinant DNA technology and its applications and the next century will see the evolution of new technologies needed for bioprocessing therapeutic drugs, proteins, and enzymes generated through recombinant DNA technology. The convergence of advanced technologies in computer science, engineering and biological sciences is producing widespread opportunities for the development and growth of companies engaged in bioinformatics, medical devices, bio-energy, agriculture, biosensors and nanotechnology related applications that have yet to be developed or imagined.

Therefore, in order to prepare our undergraduate life science students to be future research biologists, we need to transform undergraduate education, according to the BIO 2010 report.[1] This will require life science majors to develop and reinforce connections between biology and other scientific disciplines so that interdisciplinary thinking and work becomes second nature. With the integration of new technologies in biological research, biology will continue to become more interdisciplinary and will present a challenge to higher institutions that are training the future workforce and scientists of the 21st century.

The U.S. is the considered the "hot bed" of biotechnology,

---

1 *National Academy of Sciences (2003) "Bio2010: Transforming undergraduate education for future research biologists" National Academies Press, Washington, DC*

with most top companies in the field based here. The investment of federal and state funds in biotechnology-related research in universities is also immense and growing. However, according to a report by *Business Week*, biotech companies are complaining that graduating students lack the technical knowledge to carry out applied research in areas that straddle engineering, math, and computers and have little awareness of what the Food & Drug Administration is looking for when it considers whether or not to approve a drug.[2]

The vision—and therefore the challenge—in developing a new program was not only to provide students with hands-on experience with some of the widely used techniques in the field of biotechnology but also to empower them to use these techniques as tools of discovery for future research.

Addressing this challenge was the top priority for the University of Houston's (UH) College of Technology, given its proximity to the Texas Medical Center which encompasses the highest density of universities, clinical facilities, biomedical research facilities and health care institutions in the world. With 42 member institutions, the Texas Medical Center is a hub for cutting edge biomedical research. Among these institutions, M.D. Anderson Cancer research center is a top ranked cancer hospital conducting research that has significant commercial and technological value, and the University of Texas Health Science Center, which is nationally recognized as a leader in biosecurity and public health preparedness. Also located nearby is Rice University, one of the leaders in nanotechnology research.

EVOLUTION OF THE NEW PROGRAM

Traditionally, biotechnology degree programs have been housed in the university colleges of natural sciences. In our case, it was the College of Technology that took the lead in developing the academic bachelor's program. A new faculty member was hired to collaborate with industry and academic experts and procure funds

---

2   Saminather, N. "Biotech's Beef" BusinessWeek, November 6, 2006.

from state and federal agencies to develop the new program. Two advisory committees were formed: an industry advisory committee to advise on industry relevant issues, and an academic advisory committee to ensure compliance with the University academic rules and regulations. The formal process of degree approval for the program began concurrently. The program was housed in the Engineering Technology Department and the new degree plan was submitted to the college undergraduate committee.

Two curriculum tracks, one in bioprocessing and the other in bioinformatics were developed to give students the flexibility to tailor their degree based on their interests, educational background and career goals. These tracks, in combination with core courses, were intended to provide our students broad exposure to the field of biotechnology to enhance employment opportunities in the biotechnology industry and also to expose life science majors to principles and concepts in mathematics, engineering, and the use of computers in the acquisition and processing of data. In addition, the capstone course, *BTEC 4350*, can either be an internship at the local biotech industry or a research project with the honor's college. The plan is to add additional tracks as the program grows.

THE INNOVATION FACTOR: INTEGRATION OF RESEARCH AND EDUCATION

The program itself has serendipitous roots. A chance meeting of Rupa Iyer with Melinda Wales, affiliated with Texas A&M University and Reactive Surfaces of Austin, led to discussions about developing a project-based lab curriculum for the new biotechnology program. While the idea of project based curricula is not new, in this case the plan was to take the students right from the process of scientific discovery to its application in the real world and follow the life cycle of a typical biotechnology product. Laboratory exercises that covered discovery, molecular isolation techniques, transformation, gene cloning, and gene product bioprocessing were discussed to infuse 20 years of research into undergraduate labs.

The soil bacterium *Pseudomonas dimunita* is a model for this project-based curriculum. A plasmid encoded gene in this bacterium is responsible for degradation of pesticides, namely organophosphorous compounds (OP), and is activated only by the presence of pesticides in the soil, thereby providing a unique method of detection of pesticides. The OP system was chosen for this project because of the ease with which it can be integrated into the undergraduate curriculum. The bacteria are commonly found in soil and are fairly easy to isolate and grow. The OP degrading gene has been identified, cloned, and expressed, and the upstream and downstream processes of the protein's production are very well characterized.

We felt that this set of experiences was relevant and timely because of the potential application to environmental biotechnology, biosecurity, biosensors, and nanobiotechnolgy, making it extremely valuable to undergraduate curricula as students can relate the potential values of scientific discoveries in everyday life. In addition, we wanted this laboratory curriculum to integrate STEM concepts by providing training through a variety of field, bench, and computer-related opportunities that help students develop the experience and thinking skills necessary to compete in the increasingly global, interdisciplinary, and technologically complex environments associated with real world research.

DESIGNING PROJECT-BASED LABORATORY: INFUSING CUTTING EDGE RESEARCH INTO UNDERGRADUATE LABORATORIES

Three sets of activity modules that would cover topics from microbiology techniques, molecular techniques and applications to biomanufacturing techniques/technology and introduction to nanotechnology were designed. A brief description of the modules is given below. Since the lab curriculum was designed in modular format, it becomes very flexible and can be integrated in molecular biology, microbiology, genetics, environmental biology and bioprocessing lab curricula.

Module 1: Environmental Sampling and Enumeration
      Part A Sample Collection
      Part B Characterization of Microbial Community
      Part C Isolation and Characterization of Isolates
      Part D Antibiotic Resistance

Module 2: Remediation of Organophosphorus (OP) Pesticides
      Exercise 1 - Creation of a plasmid library
- Isolation of plasmid DNA
- Restriction digestion
- Ligation and transformation

      Exercise 2 - Analysis of the library
- Selection and Screening
- Screening by PCR (or restriction digest)
- Screening by Activity

      Exercise 3 - Expression and partial purification of OPH
- Expression and harvesting cell culture
- Partial purification

      Exercise 4 - Enzymatic Assay of Organophosphorus Hydrolase (OPH)
      Exercise 5 - Structure and Kinetic Characteristics of Organophosphorus Hydrolase

Module 3: Bioprocessing
      Part A Good Manufacturing Practices
      Part B Cell Growth
      Part C Bioreactors
      Part D Upstream Processing
      Part E Downstream Processing
      Part F Introduction to Nanotechnology
      Part G Biosensors

CONNECTION TO INDUSTRY RELEVANT ISSUES- THE REGULATORY AND COMPLIANCE COURSES

Industry-relevant issues were taken into consideration when developing new biotechnology courses, particularly *Biotechnology Regulatory*

Environment, Current Good Manufacturing Practices, Quality Assurance Quality Control, and *Principle of Bioprocessing.* A curriculum development subcommittee, which consisted of members from the Industry Advisory Committee, was formed to advise on the course content and assessment to develop these courses. Currently, *Biotechnology Regulatory Environment,* the very first course developed in collaboration with industry experts, is being offered in hybrid format. The course content is available on-line and students meet and discuss the material each week with an industry expert familiar with that material.

THE NEW BIOTECHNOLOGY BACHELOR'S IN SCIENCE CURRICULUM

Bachelor of Science University Core Curriculum
42 semester hours (SH)

Major Requirements – Biotechnology

> *BCHS 3304, 3201. General Biochemistry I, Laboratory*
> *BTEC 2300. Principles of Biotechnology*
> *BIOL 1362, 1162. Introduction to Biological Science, Laboratory*
> *BIOL 2333, 2133. Elementary Microbiology, Laboratory*
> *BIOL 3301. Genetics*
> *BIOL 4320. Molecular Biology*
> **BTEC 2320. Biotechnology Regulatory Environment*
> **BTEC 2321. Good Manufacturing Practices*
> **BTEC 3100. Instrumentation and Measurement Laboratory*
> **BTEC 3301. Principles of Genomics/Proteomics and Bioinformatics*
> **BTEC 4350. Biotechnology Capstone Experience*
> *CHEM 1332, 1112. Fundamentals of Chemistry, Laboratory*
> *CHEM 3331, 3321. Fundamentals of Organic Chemistry, Laboratory*

College and Department Requirements

> *ELET 2300. Introduction to C ++ Language Programming*
> *ITEC 2334. Information Systems applications*

PHIL 3354. Medical Ethics
TELS 3340. Organizational Leadership and Supervision or
HDCS 3300. Organizational Decisions in Technology
TELS 3363. Technical Communications

Technology and Other Requirements

Mathematics (10 SH)

Students are required to have credit for College Algebra through the Math Placement Exam, CLEP, or completion of course.

MATH 1330. Pre-calculus
MATH 1431. Calculus I
TMTH 3360. Applied Technical Statistics or *PSYC 330. Introduction to Psychological Statistics*

Natural Sciences (12 SH which includes university core)

BIOL 1361, 1161. Introduction to Biological Science, Laboratory
CHEM 1331, 1111. Fundamentals of Chemistry, Laboratory
PHYS 1301, 1101. Introductory General Physics I, Laboratory

Social Sciences

(3 SH) selected from core approved list.

COMM 1302. Introduction to Communication Theory, preferred

Humanities

PHIL 1305. Ethics

Biomanufacturing Track (13 SH Minimum)

BIOL 4319. Microbial Genetics
*BTEC 3320. Introduction to Quality Control/Quality Assurance
*BTEC 4101. Principles of Bioprocessing Laboratory
*BTEC 4301. Principles of Bioprocessing
Approved Elective (3 SH)

Bioinformatics Track (12 SH Minimum)

*BTEC 4300. Principles of Bioinformatics

ITEC 3343. Information Systems Analysis and Design
ITEC 3365. Database Management
Approved Elective (3 SH)

Approved Electives
       BCHS 4306. Nucleic Acid
       BIOL 4323. Immunology
       BIOL 4374. Cell Biology
       TELS 4350. Industrial and Environmental Safety
★ New Biotech courses

NEW BIOTECHNOLOGY COURSES

BTEC 2300 Principles of Biotechnology: Study of techniques and applications of recombinant DNA technology. Also examines ethical issues concerning biotechnology

BTEC 2320 Biotechnology Regulatory Environment: Overview of FDA and other regulatory agencies, steps in the approval processes, and role of government oversight and regulation during the discovery, development, and manufacture of new biotechnology products.

BTEC 2321 Current Good Manufacturing Practices: Examines the history, rationale, purpose, and GMP requirements applicable to manufacturing, packaging and labeling, testing, and control of pharmaceutical products, and the consequences of inaction.

BTEC 3301 Instrumentation and Measurement Laboratory: Provides a hands-on experience of the techniques and instrumentation used in modern biotechnology laboratories.

BTEC 3301 Principles of Genomics/Proteomics and Bioinformatics: Overview of the fields of bioinformatics and genomics. Topics, tools, issues, and current trends in these and related fields are discussed.

BTEC 3320 Introduction to Quality Control/Assurance: Quality control techniques, quality assurance issues, and quality management methods. Quality in design and planning, quality in the constructed project, and quality in production of goods and services.

BTEC 4301 Principles of Bioinformatics: This course familiarizes students with the principles and practical application of bioinformatics tools in molecular biology and genetics.

BTEC 4300 Principles of Bioprocessing: Familiarizes students with cell culture techniques, principles of bioreactor operation and purification, documentation procedures, important tasks for clean room operations, including sanitization, sterilization, cleaning procedures, calibration, and environmental monitoring.

BTEC 4350 Biotechnology Capstone Experience: Supervised internship at a biotechnology company or an independent thesis or honor's thesis with honors college.

EVALUATION AND ASSESSMENT

Our curriculum development is proceeding with learner objectives and instructor objectives in place. We are using formative and summative assessments to assess student objectives. The following is not a complete list, but rather an example of outcomes that will result from this project.

Students should be able to:
- Demonstrate their ability to reason both inductively and deductively with experimental information and data
- Explain the theory and practice of recombinant DNA technology
- Describe biocatalysis, bioprocess control, upstream and downstream processing
- Apply concepts of biology, chemistry, mathematics and engineering procedures to the spectrum of

fields making use of biotechnology including nanobiotechnolgy
- Integrate collaborative and investigative learning to build critical thinking skills
- Consider the interplay between scientific discovery and society, including the importance of scientific method and ethics

The project also has instructor centered goals that include:
- Developing a project based curriculum that integrates new technological advances into biotechnology curriculum
- Designing and disseminating laboratory activities that can be integrated into appropriate curricula
- Providing an instructors manual that will guide in implementing this curricula

Metrics for the program will reflect student learning objectives and outcomes. Each course will have a specific set of learning goals. Assessment tools will be developed in collaboration with industry advisors and assessment experts to ensure appropriate evaluation of learning is taking place. This process will include creation of exams for base-level comprehension of concepts, as well as more complex assessments (including project rubrics) for gauging higher order thinking and procedural knowledge. The specific assessments and evaluation strategies will vary depending on their applicability and utility for particular classes.

ADDRESSING LOCAL ECONOMIC DEVELOPMENT NEEDS

In addition to the academic program, we are also developing training programs in the form of short courses and post-baccalaureate certificate programs to upgrade skills and information. These are currently being developed in collaboration with local industry experts, and the plan is to offer them as continuing education courses. These will provide a strong linkage with industry and an ongoing means of communication to keep up with the rapid changes in the biotechnology and life science industry. The College of Technology also houses the Center for Life Sciences technology that supports the program in conducting outreach efforts by supporting middle and high school teacher training and recruiting students into the new program.

FUNDING AND INTERNAL SUPPORT

The program is currently funded with the help of two grants: "Bridges to the Future: Initiating a Comprehensive Biotechnology Program at the University of Houston" for $1,022,336.00 from 8/31/06-8/31/08, from the Texas Workforce Commission, and "From Nature to Lab to Production: Infusing Cutting Edge Research into Undergraduate Biotechnology Curriculum" for $121,880.00 from 12/1/06-12/1/08, from the National Science Foundation. The TWC grant has provided seed funding for equipment, outreach activities including high school teacher workshops and development of the academic training courses. NSF CCLI phase I funding is supporting the development of new lab modules and updating a *Principles of Biotechnology* course.

The biology department has provided us with lab space that will be a temporary facility for the new molecular and bioprocessing lab. Members from the local biotech industry are donating time and equipment to design a modular bioprocessing lab. Plans are also in place to have a new bioprocessing facility that will be a core facility for teaching, research, and training.

BROADER IMPACT

The new program is interdisciplinary, offers an experimental approach to learning, and simultaneously provides hands-on experience in skills relevant to the changing world of biotechnology. The program seamlessly combines college-level instruction and hands-on laboratory learning with direct relevance to the biotechnology industry.

Courses such as *Biotech Regulatory Environment, Current Good Manufacturing Practices,* and *Quality Assurance Quality Control* are developed in a hybrid format where the content is available online and students meet once a week for lectures from industry personnel who are currently involved in the types of research or procedures they will be touching upon in class. We felt this was important because it brings real world perspective to our students. These courses will also be delivered to our other campuses live via interactive television, extending the program beyond the main UH campus.

A lab curriculum that fits not only in biotechnology courses but in other biology courses can be transferable to any institution. Within a year of its development, the project-based lab curriculum is already being adopted and beta tested by Purdue University and Brigham Young University. Organophosphorous compounds are used globally and we are currently discussing collaborations with academic institutions in India and New Zealand to replicate the program globally.

Distance Education
Richard Conroy

Richard Conroy, Ph.D., is an Adjunct Assistant Professor at the University of Maryland, University College. Richard may be contacted at *rscus2001@gmail.com.*

HISTORY OF DISTANCE EDUCATION

The delivery of education to students who are not "on campus" has been a challenge facing institutions for more than a century. The earliest correspondence courses relied on the development of an efficient and cost effective communication channel: the postal service. Isaac Pitman realized in the 1840s that the penny post system in the United Kingdom was a powerful medium for the dissemination of his shorthand method. Within two decades the University of London was the first degree-granting institution to get involved in distance education, recognizing the geographic spread of the British Empire need not be a hindrance to gaining a degree. It founded its "external programme" in 1858, a program that has since diversified to over 100 different courses with 40,000 students in 180 countries.

Following the Second World War and the widespread adoption of radio and television, a new wave of distance education programs aimed at adult and non-traditional learners was developed to take advantage of these distribution channels. Among the first of these was the Open University in the United Kingdom, founded in 1969, which became a model for many similar institutions around the world. In addition to using the postal system, these higher education bodies began to distribute course content via the phone, radio, and television, taking advantage of off-peak hours on public channels whose mission included supporting continuing public education.

The advent of tape recorders, video cassette recorders, compact discs, and personal computers in the last three decades of the twentieth century has diversified the channels through which educational content can be distributed yet further. However, with the exception of the slow postal service and the historically expensive telephone network, there has been limited opportunity for real-time teacher-student interactions, nevermind peer-peer interactions. The development of the publicly accessible Internet a decade ago has removed this barrier and opened another chapter in the history of distance education, enabling an educational experience closer than ever before to the "on campus" experience.

The dotcom boom of the late 1990s saw an explosion of interest in using the Internet to deliver education courses. However, most of these companies have since vanished or targeted alternative markets. These businesses failed for many different specific reasons, but in general they underestimated the challenges and over-estimated the demand. For example, Arizona Learning Systems was created in 1996 and was provided with $3.8 million by the state legislature to build a distance education program. However in its eighteen months of operation it attracted only 118 students.[1] Startups were not alone in underestimating the difficulties; the U.K.'s Open University, with more than twenty years experience in distance education, cancelled its U.S. online operation in 2002 after recruiting only 660 registrations, despite $20 million of investment.[2] Perhaps the most damaging failure during this surge was the collapse of the for-profit Masters Institute (San Jose) in a storm of fraud allegations.[3]

Arguably, one of the most successful online ventures is run by the non-profit University of Maryland, University College

---

1   Carnevale, D. (2002a). *Arizona lawmakers want to sell computers from failed distance program. The Chronicle of Higher Education, 49(05), NA.*

2   Arnone, M. (2002). *United states Open U. To close after spending $20-million. Chronicle of Higher Education, 48(23), A44.*

3   Carnevale, D. (2002b). *Questions linger over rise and fall of online program. Chronicle of Higher Education, 48(21), A27.*

(UMUC).[4] UMUC has successfully been able to distill the critical elements and experience from its established core distance education business to realize and expand demand online. Starting in 1947 as an adult education university, UMUC first offered on-line courses in 1996 and the first complete online degree program in 1999. The Graduate School of Management and Technology, which the biotechnology program is part of, had more than 115,000 credit hour enrollments in 2007 in fourteen degree programs, all online, with healthy annual growth.

The success of a number of institutions, the continued expansion of Internet access and its ever-increasing usage, the saturation of traditional student markets, and demand from students to offer more flexible content delivery has encouraged many higher education establishments to cautiously re-enter the online distance education market. The rapidly increasing range of tools and diversity of integrated teaching platforms has nurtured this expansion along with the desire to tap the profitable and expanding adult and non-traditional learner market. The brisk and sustained growth of the biotechnology industry and its demand for an educated workforce is an ideal segment to target, particular for solely online on mixed-mode delivery.

CURRENT TRENDS IN ONLINE LEARNING

The Internet has had a radical impact on society. In 2002 the Pew Trust reported that more than three quarters of high-school students regularly use the Internet for research.[5] In higher education, a 2005 Sloan Consortium report noted that nearly two-thirds of institutes offer at least one online course and more than half offer at least one blended course, with medium to large, publicly funded institutions taking the lead.[6]

4   http://www.umuc.edu/

5   Minkel, W. (2002). Pew study: Students prefer 'virtual library'. School Library Journal, 48(10), 28.

6   Allen, I. E., Seaman, J., & Garrett, R. (2005). Blending in: The extent and promise of blended education in the United States: The Sloan Consortium, Needham, MA.

Data from the National Center for Education Statistics show that a third of all institutions now offer a complete degree or certificate program online—more than 2,800 programs—with the majority being four-year undergraduate degrees.[7] Altogether this equates to nearly 120,000 different courses being offered online, the majority of institutions have fewer than 500 students for their online offerings, and many courses are designed as copies of face-to-face classes. Business and liberal arts programs have seen the fastest adoption of online and blended courses, while social sciences and history are lagging behind, though the gap is decreasing.

In their 2006 survey, the Sloan Consortium reported that 3.2 million students took online courses in 2005, an increase of 800,000 from the previous year, with no evidence of the rate of growth decreasing.[8] The sustained growth in the market over the past five years has been noticed, with approximately a third of institutions planning to start or increase their number of online or blended course offerings.

The composition of students taking these courses mirrors the general higher education body, with slightly more graduate-level students and a focus toward associate degrees and larger institutions. Perhaps surprisingly, the majority of distance education students live within one hour of the nearest campus of their institution, suggesting that flexibility of study time is a major factor in choosing online education.

There is little difference, on average, with how face-to-face, blended, and online courses are staffed within an institution, because core faculty often teach the same course in different settings. However there are significant differences between institutions, with some solely using core faculty while others use up to 90 percent adjunct faculty. The National Education Association reported in 2000 that distance education faculty prefer web-based courses

7   *Waits, T., & Lewis, L. (2003). Distance education at degree-granting postsecondary institutions: 2000-2001: U.S. Department of Education, National Center for Education Statistics, Washington, DC.*

8   *Allen, I. E., & Seaman, J. (2006). Making the grade: Online education in the United States, 2006: The Sloan Consortium, Needham, MA.*

to video- or book-based courses, with a good correlation between the degree of faculty-student interactions and faculty satisfaction with the course.[9] The majority of faculty reported that distance education courses are more time consuming than their on-campus equivalents, however this is not taken into account in pay or duties. Online faculty also report dissatisfaction with compensation for intellectual property they generate and a lack of acknowledgement of "non-contact" hours.

Contrary to the popular impressions, most institutions cap the number of students that can enroll in an online class, with more than two thirds of classes having less than 40 students. Based on this treatment of on-campus and distance education classes being equivalent, the majority of educational establishments (77 percent) have the same pricing structure.

The major disincentives starting online courses, reported by the third of degree-granting establishments not planning to offer distance education programs within the next three years, are a lack of fit with the institute's mission and program development costs. Of the approximately 150 graduate-level biotechnology degree programs in the United States, fewer than ten can be completed solely online, with one of the most visible and long-running being UMUC's Master of Science in Biotechnology Studies. The biotechnology program at UMUC became a stand-alone in 2001 and has seen growth to more than 350 enrolled students per semester. The small but growing number of online biotechnology courses highlights the challenges with delivering quality, diverse, science-based distance education relative to on-campus courses. However through careful planning, the judicious use of the Internet these challenges can be overcome to address the growing pool of people seeking employment in the biotechnology sector.

---

9   N.E.A. (2000). A survey of traditional and distance learning higher education members (Vol. NCES 98-062): National Education Association, Washington, D.C.

DELIVERY METHODS

Today the choice of content delivery methods is determined by the target audiences and core competences and mission of the educational institute. The audience is increasingly technically literate with the vast majority of higher-education courses now including computer-enhanced instruction, making use of the interactive and multimedia capabilities of computers and the information capacity of the Internet. The increase in the number of higher-education institutes over the past two decades has required many of them to more carefully define and sustain their position and unique advantages, as well as to diversify into online education.

A helpful metric for describing course offerings is the percentage of the content delivered by computer or online: courses with no online or computer-assisted component are referred to as "traditional" or "face-to-face," while those using up to 25 percent online or computer-assisted components are "technology enhanced"; 25-75 percent are "blended" or "hybrid" courses; and greater than 75 percent considered to be "online" or "computer-based" education.

While computers and the Internet now dominate distance education, the University of London's external program continues to build on its textbook-based correspondence course model, which is compatible with reaching its base of students in developing countries where technology and Internet access are limited. These courses are primarily aimed at professional graduates who are capable of working independently and offline from their peers, though textbooks and readers are supplemented by optional online forums. Proctored exams are taken at recognized testing centers, completing this traditional, paper-rich approach to distance education.

At the other end of the spectrum, traditional face-to-face universities such as Harvard University Extension School, record their classroom lectures for distance education purposes and provide them on the Internet for access by distance-education students. These videos are supplemented by electronic files of the presentations, homework, and exams, which can be submitted or taken online. This adoption of on-campus education styles to a distance learning environment clearly has some advantages, though it often lacks the faculty and peer interactions found in dedicated online programs.

Many approaches exist between these two distance education models. The challenge of disseminating large quantities of information to students who may have low speed or intermittent Internet connections has prompted some educational institutes to take a hybrid approach to distributing their course material using other formats such as CDs and DVDs, supplemented by online forums. Similarly, open universities, founded before the Internet, continue to build on their content delivery through television, radio broadcasts, and pre-recorded videos, supplemented with online interactions. Using pre-recorded content provides a level of convenience to students, who do not need to have a permanent access to the Internet or a computer. However, this approach is more rigid and requires an established course structure and much more infrastructure to support. Informal, non-accredited, distance education organizations, such as The Teaching Company, have also reached out through pre-recorded media with books, audiotapes, and videos available in many bookstores and libraries.

An alternative approach has come from institutions, such as University of Maryland University College, which specialized in distance education before the advent of the Internet. These institutions have adopted their methods and materials for asynchronous delivery through an online teaching platform integrated with other university services. This seamless, online approach improves consistency between courses and interactions online but has the disadvantage of requiring an Internet connection to access the classroom.

A third hybrid model employed by a number of institutions is the concept of a low-residency program, where distance education courses are supplemented by a number of intense face-to-face sessions. This model increases student interaction and retention, though it discourages the enrollment of students who cannot make the time commitments to attend the residences.

In addition to these distance education models, traditional universities have started blending face-to-face and computer-based education. This is a growing but discrete transition, which may not be designed or aimed at distance education. Instead, it is generally used to enhance the experience of on-campus students. Although blended courses may appear to offer the best of both models, the Sloan Corporation reported in 2005 that the rate of adoption blended course offerings has slowed and they are not regarded by academic leaders as holding as much promise as fully online courses.[10]

In devising a distance education program, the most success-full institutions have blended online delivery methods with their traditional strengths, providing an integrated learning package with minimum overheads. In their 2005 report "Blending In," the Sloan Consortium reported that students have no preference for one particular content delivery method. Instead, they are open to a variety of modes according to circumstance and sophistication.[6] This reflects the fact that a majority of course participants have now experienced an online or hybrid course in their previous education or work environment. Flexibility in delivery can enable educators to explore mixed-mode programs within the limits

---

10   Allen, I. E., & Seaman, J. (2005). *Growing by degrees: Online education in the United States, 2005: The Sloan Consortium, Needham, MA.*

Box 1: Ten Institutional Challenges

1. Over-estimating interest – low enrollment plagues online courses and is often attributed to a lack of a structured learning environment, which is required by many less-experienced learners.
2. Under-estimating resource requirements – rarely are the long term costs of running and updating hardware and software as well as personnel and training costs fully accounted for.
3. Retaining students – after finishing an online course, regular interaction with students immediately stops, making it difficult to retain students' interest. This is particularly problem-

atic over a summer break, where routines are broken and course enrollment deadlines are far in advance.

4. Duplication of effort – with multiple courses and programs being developed simultaneously, the wheel is constantly reinvented.

5. Faculty under appreciation – failure to acknowledge the increased workload associated with online classes and resolution of intellectual property can lead to increased reluctance of full-time faculty to be involved with online programs.

6. Infrastructure integration – enrollment of distance students, tracking their progress and payments, and providing feedback to them as well as providing access to library, support and administration services means that university systems need to be accessible online and integrated with the online teaching environment, with security and administrative implications.

7. Maintaining quality - rapidly changing student numbers, and the perception that learning outcomes are somewhat inferior online need to be addressed to ensure quality.

8. Limited feedback – evaluation of programs, courses, faculty and students is hampered by limited access to the student in an asynchronous, non-spontaneous environment.

9. Downtime – one of the biggest fears for any online program is a prolonged system or Internet downtime which inevitably will happen.

10. Poor alumni support – without the richness of an on-campus lifestyle and face-to-face contact between students and university faculty and administration, the lack of connection with their institution and cohort may lead to poorer alumni support.

of the access requirements and expectations of their students and faculty. The diversity of content offered in on–campus biotechnology degree programs, from lab classes to theoretical science and practical business courses, is hard to mirror completely online, and thought must be given to key strengths and student demand. At UMUC the focus is on self-guided courses which focus on professional development, and is typified by the capstone course where teams of students work with local biotech startups to develop and assess technologies and marketing opportunities.

INSTITUTIONAL CHALLENGES: PROGRAM DEVELOPMENT AND INTEGRATION

The development of an online program and courses can be a daunting task. Box 1 details ten of the most common challenges faced by institutions. For discussion purposes, the development process can be broken down into three steps: the creation and integration of the program into the institutional framework; the day-to-day running of the courses; and, revision of the program based on feedback.

A number of general strategies can streamline the creation process. Through trial and error, the best approaches articulate the goals of the program in general and courses in particular, and then choose an appropriately modified course development model to fulfill these goals. The program model should build on common educational theories such as *Bloom's taxonomy*[11] and *Gagne's nine events of instruction*[12] to systematically develop learning in the program and identify objectives for assessment. Individual courses can be adapted to follow particular models and meet course criteria, including the use of faculty-centered, student-centered, or team development models. The advantage of establishing a common template for courses, identifying and archiving reusable resources, creating learning objectives to balance instructional activities, and developing only one course at a time cannot be underestimated at the planning stage. For a new program, the course creation phase is time intensive and can take up to one year, but careful develop-

---

11 Anderson, L. W., Krathwohl, D. R., Airasian, P. W., Cruikshank, K. A., Mayer, R. E., Pintrich, P. R., et al. (2000). *A taxonomy for learning, teaching, and assessing — a revision of bloom's taxonomy of educational objectives (Second ed.).* Boston, MA: Allyn & Bacon.

12 Gagne, R. M., Wager, W. W., Golas, K., & Keller, J. M. (2004). *Principles of instructional design (Fifth ed.): Wadsworth Publishing.*

ment will decrease workload during the teaching phase.

Increasingly, courses are built on a number of technology platforms designed to host online distance education programs at the enterprise level. These platforms are often referred to as learning management systems (LMS), which organize content, support different learning and testing tools, and integrate with other institutional systems. Amongst the most common are WebCT & Blackboard, eCollege and Desire2Learn. In addition, there are a number of other platforms such as the open source Moodle, the proprietary rEsource used by University of Phoenix, and WebTycho used by UMUC[13]. For students, all of these platforms can be accessed using any type of web browser, with authentication, online assistance, access to their grades and seamless integration of multiple classes. For faculty, there are a wide range of features for course development and management, including multimedia integration, threaded discussions, testing and grade history, as well as online training and staff development. For institutions, these platforms can seamlessly integrate with other resources and databases, providing tracking and documentation of student status and performance, linking to web resources such as library databases, and administrative oversight. For a more detailed comparison of there learning management systems there are a number of reviews available on the Internet.[14]

Typically four people are involved in a course creation team: the program manager who identifies course needs and ensures integration with program goals, two faculty members who develop course structure and content, and a technology specialist who helps integrate a existing templates and resources into the course as well as creating new content delivery. Institutional support from the beginning is critical to avoid isolation, duplication and unmanageable workloads for the faculty, and to provide integration with resource and management systems. Institutions also need to clearly define responsibilities and expectations, particularly with

13 http://webtycho.umuc.edu/
14 http://kumu.brocku.ca/webct/LMS_Options_and_Comparisons
http://www.edutools.info/static.jsp?pj=8&page=HOWTO

respect to administrative tools and how online education fits into their mission. Before program development begins, advice is often sought from local, state, and national education consortia which can provide invaluable advice and guidance.

The program manager is typically tasked with integrating elements of the institution's mission, infrastructure, and technology base into the course and program, often with assistance from other administrative, technological, and teaching staff. Other tasks include preparing projected resource requirements, student numbers, and staff levels. Recruiting a diversified faculty pool and advertisement of the program are ongoing concerns. The program manager also ensures that policies are established for staff and student assessment, acknowledgement, and compensation.

As part of the creation team, faculty members are involved with the refinement of learning objectives, creation of course content, development of specific instructions and rubrics for each graded assignment, identification and integration of textbooks and reading materials, establishing a grading system, and building a syllabus. Faculty should also be aware that workload balance can be achieved in course design through limiting the number of graded assignments, using peer and self assessment, and self-guided assignments. Courses must be designed to be adjustable according to semester dates, unforeseen access problems, and over- or underestimation of student ability.

The role of the technology specialist in the creation team is to provide expert input and continuity for online classroom design. Perhaps the biggest component of the specialist's work will be in creating focused multi-media content for the course in conjunction with the faculty, and ensuring compatibility. On an ongoing basis, the specialist can suggest and implement labor saving ideas like placing dynamically changing course elements, such as Internet links, in a separate area, or finding alternative means to create a more permanent way of referencing them. Existing course templates should be introduced and used by the team where possible to provide style continuity. In addition to course content, adequate library, technical, and administrative tools must be available to faculty and students online.

During the teaching phase, day-to-day oversight of course content and student interactions is the primary responsibility of the teaching faculty. In order to keep the program running smoothly communication channels need to be sustained between the students, faculty, and administration. Between the administration and students, online access to support and financial services needs to be easily available and straightforward, with a defined quality of service. Within the online environment, students should be encouraged and supported to create and participate in clubs and societies and in the governance of the institute, and in alumni clubs after completing their programs. Between the administration and faculty, easy access to personal development courses, course management tools, support services, regulations and policies, and the governance structure are also important, as well as scheduling of future teaching duties and compensation. Between faculty and students, online "office hours," guidance on etiquette, grading organization, and participation in events outside the online environment improve communication and awareness. In addition to these interactions between the shareholders, the hardware and software channels supporting these interactions must be maintained by the technology specialists.

One of the biggest challenges with online courses is soliciting meaningful feedback for developing and restructuring courses and programs. The most effective feedback schemes make it mandatory for both faculty and students to complete online survey forms before the course can officially be completed. In the case of students accessing an online classroom, one approach is to prevent them accessing the classroom area one or two weeks before the end of the semester, until they complete an online feedback form. For faculty, a survey may be included with the final grade form, requiring both to be completed before the course can be closed. These staff and student surveying methods can be supplemented by less frequent surveys of other stakeholders like industry partners, collaborating institutions, and professional bodies. More in-depth periodic reviews using external experts should be carried out to ensure quality and confidence, particularly for newly established programs and upcoming accreditation reviews.

The biotechnology program at UMUC was recently restructured in response to this type of broad collection of feedback, moving from a single program with electives, to a structure offer-

---

Box 2: Ten Faculty Challenges

1. Open all hours – online education can turn into a 24 hour job when juggled with other commitments.
2. Lack of interaction – failure to stimulate students can lead to a decrease in motivation and classroom responsibility.
3. Compensation – lack of structured agreements for teaching online courses has been unsettling for face-to-face educators pushed towards the online arena.
4. Plagiarism – working in an online environment provides a greater temptation to copy and paste information with proper acknowledgement.
5. Testing – enforcing conditions for online testing is problematic and can be addressed by either requiring end of semester exams to be taken in-person at a testing center, or for exams to be "open book, take home" exams.
6. Group projects – implementing group projects online is challenging; the asynchronous nature of distance education makes it difficult for groups to discuss projects in real-time and develop a cohesive plan.
7. Losing control – not having the class sit in front of you, shifts the learning focus to the students and makes informal feedback and assessment of progress difficult.
8. Approaches to learning – some students and faculty prefer instructor-centered learning in a structured learning environment.
9. Technophobia – with the rapidly changing ICT world, time and effort spent learning one package can become quickly redundant, increasing faculty resistance to the adoption of new technology.
10. Isolation – adjunct faculty not teaching at on-campus develop feelings of isolation from their peers and the administration.

ing three specializations: bioinformatics, biotechnology management, and biosecurity and biodefense. These specializations arose organically from the existing program structure and the interests of faculty, students and industry, making the transition to the new program structure easier to implement. The lab course, based on an intense full-time face-to-face short residency was dropped from the program because of the high administration overhead, uncertainty in predicting student numbers and lack of fit with the three streams.

One unique but common issue faced during any restructuring of an online program is that although students begin at the same time, some may finish after only one year while it may take others several years. In implementing a program, developers have the option of requiring all students to proceed at the same rate, or to give them the flexibility to take courses as their schedule allows, though the latter requires a long grandfathering period before any changes can be fully implemented, and can produce administrative headaches.

FACULTY CHALLENGES: CLASSROOM MANAGEMENT AND CAREER DEVELOPMENT

The consistent double-digit growth of online education over the past five years has been met with mixed emotions by faculty. From concerns over job security and teaching standards through to the desire to experiment with new teaching methods and technologies, educators have embraced online distance learning with some reservations. Box 2 lists some of the key challenges faced by online teaching faculty, which need to be addressed in a successful online program.

This diversity in faculty opinion mirrors institutional attitudes. At one extreme, universities have adopted an open-source attitude and full acceptance of online education, perhaps best typified by MIT's OpenCourseware project which shares more than 1700 courses online. At the other end of the spectrum, establishments such as Yale University do not currently offer much online beyond lecture videos. Faculty's perception of their role is driven by institutional attitudes and expectations and, like any workforce facing upheaval, must be maneuvered gently to implement the changes brought by online teaching.

Some of the resistance to moving courses online is rooted in a number of concerns related to working conditions and compensation. The rapid growth of for-profit institutions employing adjunct faculty has added to the impression that permanent tenured positions are being replaced by on-demand, part-time adjunct positions. The lack of structured compensation agreements and acknowledgement of the changes in working practice required by online classes has compounded this problem. Institutional policy—or the lack of it—can drive faculty perception either way very easily, and the political implications of developing online education programs must be carefully managed.

Educators face a number of unique challenges within the online classroom. Traditional instructor-centeric teaching models, where the focus and control of the class lies with the faculty member, are difficult to implement online, which lends itself more naturally to student-centered learning. The shift in conscious responsibility for learning from instructors to students feels like a loss of control to some faculty with extensive face-to-face teaching experience. Often this shift in perception is described as a transition from an "instructor" in a face-to-face classroom to a "facilitator" in an online classroom. The importance of training, organization, and feedback cannot be emphasized enough in helping faculty make this transition smoothly. Mentoring from established online educators, staff development courses based on the same technologies used in the classroom, and an emphasis on clear communication and the collection and use of focused feedback can help allay fears. Good habits to establish in the classroom as part of this staff development include preparing and publishing syllabi in advance which include details on grading schemes, assignment schedule, participation expectations, how to contact the instructors, and the speed and nature of feedback students can expect. Ensuring that students and staff have sufficient technical skills to participate in the class can be achieved by providing introduction courses and access to syllabi before the beginning of teaching. Having prepared participants and course materials in place before the start of the semester will minimize timing overheads and maximize the online experience for all involved. At UMUC, the Center for Support of Instruction produces an electronic resource for faculty, DE Oracle,[15] and the Center for Teaching and Learning organizes professional development courses[16].

The anonymity of online participation and ease with which information can be accessed on the Internet is a major cause of faculty concern, with many institutions actively encouraging the use of identity verification through testing centers and plagiarism detection via services such as *turnitin.com*. Awareness among faculty of how to detect and deal with dishonesty, combined with robust institutional level systems can minimize the friction caused by these practices. While in-person testing may be inconvenient and expensive for the students and create administrative overhead, because testing centers are generally only located in large cities and may require long commutes over multiple nonconsecutive days, they do provide an important service to educational institutions at a number of levels and should be considered at both graduate and undergraduate level. Mixing formal assignments with informal discussions in online classes provides a less concrete form of verification at the class level along with the constant modification of assignments, multi-part assignments with submitted drafts, and a robust honor code. Differences in quality and writing styles between assignments can also be picked up by instructors and may provide grounds for more careful scrutiny, though without program-level oversight, variation in performance, and quality between classes cannot easily be detected. Plagiarism remains a constant threat to integrity in the classroom and the early education of students, active and obvious use of plagiarism detection services and a robust policy can help minimize the extent and impact of incidents if and when they occur.

---

15 http://deoracle.org/
16 http://www.umuc.edu/distance/odell/ctla/index.shtml

Feedback, modification, and development both during and after the class are important tools for faculty to master online. The ability to adjust and modify new or difficult tasks, such as group projects, is a crucial skill particularly when student feedback is used in performance assessment. Feedback at multiple levels should be used as part of the evaluation of students, faculty, and administrators, as well as for course and program restructuring. As part of this process, students should be acknowledged for providing feedback. To assist other faculty, a method for recording feedback and changes made in courses and programs should be implemented in the form of a history document, updated after the end of each semester to provide notes, guidance, and insight.

As the linchpin in online education, faculty fears have to be addressed and their teaching instincts engaged for a program to be successful. At UMUC, experience has shown that good management can prevent problems snowballing, and faculty need to be actively engaged with incentives and the right tools to enthusiastically participate in online learning.

STUDENT CHALLENGES: GETTING STARTED ONLINE

The flexibility of distance learning appeals to a much broader group of potential students than on-campus education. Falling into this group are adult learners—a group with complicated time commitments because of employment and family—who are seeking to improve their education for tangible career benefits. This group can also include people looking to change professions, receive formal training in a pastime or interest, or simply seeking intellectual stimulation. The second group not fully tapped by on-campus programs is "non-traditional" students: people who have life experience, but not necessarily formal qualifications. These students may come from a vocational background, a field which values apprenticeship, organized into guilds, or from service overseas. Online learning enables these two groups of students to connect with programs which meet their needs, removing geographical and time pressures which may prohibit them attending an on-campus equivalent. Targeting these groups through active outreach can be difficult, but can also lead to a sustained flow of students through word of mouth in these often tight-knit communities.

The challenge of learning online cannot be underestimated. Faced with the option between a distance education and a face-to-face section of the same class, some students will choose the distance education class because they perceive it as saving time and enabling them to pursue other activities. While they can choose their hours and there is no commuting involved, distance education classes require at least the same amount of study time. For online classes, interactions have to be typed and read, slowing communication of ideas, and learning online is asynchronous, student-centered and isolated, which can slow absorption and understanding.

In a traditional three credit face-to-face course, two and a half hours per week are typically spent in a classroom, with students expected to spend at least the same amount of time studying and preparing for class. The time-consuming nature of online classes can be illustrated by considering an online class of thirty students, with each student posting ten substantial messages each week, where it can easily take three hours alone to read through these messages and provide responses, before time is spent formulating your own responses to the discussion topics. Inexperienced students can easily overextend themselves by signing up for too many distance education classes, without realizing the time commitment involved, and that instructors will be looking for significant contributions and sustained participation.

Box 3 describes ten best practices to encourage in online students. Perhaps the most important skill to instill in online students is the ability to work independently. Self-motivation, leading to good time management, budgeting and active communication is critical to achieving the highest grades.

At an early stage before classes begin, technology difficulties need to be ironed out through access to responsive support services

Box 3: Ten Student Best Practices

1. Take responsibility – it is easy to ignore online courses, because there is no regular schedule and the responsibility for learning is on the student. Create a work plan, devoting time each week to complete assignments well before any deadlines.
2. Have the right equipment – a reliable computer, Internet access and word processing package are the minimum requirements for an online course, but familiarity with a wide range of ICT concepts packages is highly desirable.
3. Read the syllabus and course materials – be proactive in understanding the demands on the instructor and the format of the course. Learn to navigate the online classroom and be aware of where and when assignments need to be submitted and exams taken.
4. Be interactive – participation is graded in online classrooms,

and self-guided tutorials designed to familiarize the student with the resources and templates used in the program. Part of this initial training must include making the student aware that they will be spending a lot of time in front of a computer and they will be expected to check classroom activity a minimum number of times per week. An introductory course covering basic skills, such as using library resources and writing scientific papers, can also be used to introduce good working practices, participation and assignment requirements and expectations for the rest of the program.

Visible, value-adding, contributions are important for making a student's presence felt online. In a face-to-face class implicit participation can be observed by the instructor, even if the student is hiding at the back of the class and avoiding eye contact. But online contributions have to be explicit and significant, putting more pressure on students to be interactive both with their peers and with the instructor. To ensure participation in online classes, contributions are usually assessed as part (10-25 percent) of the final grade, placing a requirement on students to be more interactive with their peers as well as the instructor. So while there is no

so be prepared to make early, substantial and sustained contributions as well as reading and responding to peers.

5. Add value – when participating online avoid being too casual, keep your messages succinct and on topic and make it easy for readers to follow the context and who you are. Stating opinions supported by references will impress the instructor much more.

6. Check material and presentation – with integrated spelling and grammar checkers, there is no reason to submit work with errors. Work should be submitted in accordance with style guidelines and properly referenced to avoid plagiarism.

7. Avoid emotional messages – with asynchronous online class-rooms you can step back from getting involved in "flame wars". Take time to think and compose dispassionate messag-es to fellow students or instructors to avoid highly emotional language or inflaming situations.

8. Use resources – access to library and support services are of-ten integrated into the teaching environment. These provide a wealth of time-saving, financially beneficial and grade-im-proving information making it worth the effort to navigate through these resources

9. Give feedback – if there are elements of the course you partic-ular enjoy, or which are above or below the level of the class inform the instructor.

10. Develop a network – studying online can be challenging because of the feeling of isolation. Friends and family can provide a local support network to help you succeed. Devel-oping online interactions outside the course can be difficult, but if they do not already exist, encourage your institution to provide virtual chat cafes, societies and alumni organizations to support students.

defined deadline for reading assignments as with a face-to-face class, a student cannot go through the course without reading the assignments and affecting their final grade. There is also a trend toward rewarding students for participating in discussions earlier, where their posts can be read by other students who are able to comment on them, compared to students who read and submit their work just before deadlines.

The choice of having daily, weekly, or monthly assigned tasks, or choosing a self-guided course where students set their own timetable is the option of the course creation team and the administration. Many students perform better in a structured environment, and the inclusion of team assignments virtually necessitates the delivery of milestones to ensure progress is being made. In an individual study course, although there may be no defined study schedule, participation can still be encouraged through informal discussion boards. Requiring a level of participation can be challenging for some students who need to project a modified personality online or have variable time-consuming commitments and can only access the classroom once a week. Whatever format is used, time management is critical in ensuring regular study of the course materials, and ensuring that major assignments are planned for. Inevitably, students who don't plan ahead are the ones who post assignments within minutes of the deadline, do not read the assignment details carefully, and begin to slip behind on reading assignments and participation. Therefore, introductory courses should introduce good time management skills and how to plan for courses.

At UMUC faculty are free to experiment with teaching methods beyond the common syllabus components. Some good practices which have been identified by the faculty include rewarding early participation in discussions, defining clear weekly participation, and providing rubrics for all graded assignments. The open access, asynchronous, student market which UMUC tries to address provides a range and diversity of students which can be challenging to instructors and students alike. Introductory courses for faculty and students are used to start everyone on the same page, and students are guided through the programs at their own pace.

THE FUTURE - WEB 2.0 AND VIRTUAL LABS

The current trend on the Internet is away from centralized, curated, stores of information, toward a widely distributed network of information published by small niche groups and individuals. The ability to aggregate this information has spawned the search giants, retailers, and social networking sites which dominate the dot-com market. The decentralization of information and the tools which are making this transition possible are also having an impact on online education, towards what some have dubbed "e-learning 2.0."[17]

This trend has been driven in part by the changing student demographic, now composed of a generation who have grown up with personal computers and Internet access. These "digital natives" are not as constrained by geographical considerations, and expect interaction and near-instantaneous responses from their peers, and go beyond simple consumption of media. The fragmentation of markets and the empowerment felt by this generation of learners has pushed the student-centered design model, with a number of educators now experimenting with the new tools of "web 2.0."

The trend of having 100 percent wireless campuses with interactive classroom participation systems to engage laptop-wielding students is already altering on-campus classroom dynamics. The development of dynamic web servers and peer-to-peer networking have helped change the dynamics of online teaching, enabling individual users to create, manipulate, and distribute information content with ease from their own computer.

BLOGS

Web logs, or "blogs," have seen sustained growth since their popularization more than a decade ago. Most blogs are chronological, textual, records created by individuals for public consumption. Building on the concept of bulletin-board systems and the Usenet news service, hardware and software advances have enabled this type of personal journalism, creating the same "long tail" phenomena for journalism that has developed in shopping and information.

A recent Technorati "state of the live web" summary reported

17 *http://elearningtech.blogspot.com/2006/02/what-is-elearning-20.html*

the existence of more than 72 million blogs with an aggregate average of 1.5 million posts per day. Perhaps surprisingly, Japanese is now the majority language for new postings, signaling faster growth in Asian countries. Education-related blogs on only one hosting service (*edublogs.org*) number more than 30,000. Looking beyond the numbers and hype, less than a quarter of blogs are active, with the active percentage and the average number of posts decreasing, suggesting that growth is slowing as users explore other technologies and outlets. There are several hundred blogs related to biotechnology, covering areas from biotech investment and healthcare management to patent reform and drug development.

Blogging has been incorporated into the pedagogy of several online classrooms, complementing the centralized hierarchical structure of learning management systems with the distributed, open, and independent nature of blogs, reinforcing a student-centered approach to teaching. One of the most common uses of blogs in the classroom is as a guide and commentary of assigned reading. As a reading diary, students post comments and critiques of the course material with the possibility of feedback from peers and teachers. In journalism classes, blogs have been used to understand how news stories spread and develop without being limited by the spatial, temporal, and access constraints of being in a classroom. For group projects and long-term individual assignments, blogs can help individuals provide a chronicle of work done and can help educate students on using this technology while providing another tool for assessment by faculty. For distance learning programs in particular, blogs have provided a level of social interaction beyond that in integrated learning platforms, providing continuity beyond the end of courses, and opportunity for independent reading, writing, and analysis.

As with the other web 2.0 technologies, there must be a clear structure and purpose for using the technology in the classroom. Using blogs as general purpose discussion boards or as supplemental learning management systems does not give enough purpose to sustain regular posting, and too much instructor oversight can dampen student expression and enthusiasm. Using the right tools for the right objectives is key to successful blogging assignments, particularly when class sizes get large. Some simple guidelines and good aggregation software can make life for the instructor much easier, and help students explore their own learning styles. Blogging has made more of an impact on liberal arts programs than biotechnology and other sciences because of the reading-, reflecting-, and responding-nature of blogs. The business and management sides of biotechnology, however, are ripe for blogging assignments and it is only a question of time before some of the more experimental online biotechnology programs begin experimenting with them.

PODCASTING

The ability to record individual audio and video files and make them available in a widely recognized format has started to push online education beyond the use of text chats and Powerpoint and Word files. Building on the widely available iPod & mp3 player formats, podcasts quickly reached the mainstream in 2002 because of their multimedia appeal, minimal production costs, and ease of distribution. Higher education has sought to make use of this technology to provide another medium, more compatible with existing learning, to reach students and connect with their daily lives. The ability to record lectures for future playback is advantageous for both face-to-face teaching and for distance education, blurring some of the distinctions between the two. Podcasts also have the advantages that they can be downloaded using peer-to-peer networking and can be locally stored for offline and mobile consumption.

A number of services for aggregating podcasts have sprung up, most notably iTunes U, from Apple Inc., creators of the iPod. While providing seamless integration with their existing iTunes store front, a number of questions have been raised in the higher education community about the use of proprietary formats through a closed access system which is not open to searches or referencing. Initiatives such as Ed-cast by the University of Illinois have sought to address some of these criticisms. Some have also argued that the rise in higher education podcasting is illusionary because many recorded lectures are now being re-branded pod- or vod-casts, and that methods for integrating this technology into mainstream classrooms are not clear.

The freedom and ease with which podcasts can be made is also having an effect on how students interact with their peers, professors, and the wider society. Presence online is becoming a growing part of life and often this means taking part in online social networks and recording events and information on the Internet. With this in mind, podcasting has been most easily integrated with creative courses where students keep diaries or complete multimedia projects. Podcasting has not had a significant impact on biotechnology programs, though with many small startups and entrepreneurs, there is ample opportunity to incorporate interview-style podcast assignments into online classrooms and for students to developing marketing projects making use of multimedia technology.

WIKIS

The concept of open databases has given rise to "wikis" or online collaborative websites for information. Perhaps the most famous of these is Wikipedia, the free online encyclopedia containing more than 2 million entries, where articles can be edited by any user.

Wikis are starting to be used in a number of classrooms, with the goal of building up collective knowledge which can perpetuate beyond the confines of the class. Wiki collaborations are not limited to students and do not have to be created specifically for the class. Assignments can look at the pros and cons of using Wikipedia for research and then focus on creating new content for it or similar wikis. Groups of educators can pull together and interact with each other through a wiki to develop and refine course materials or write textbooks. Many online educators have also found it beneficial to share their experiences through local and state distance education consortiums, providing resources for new faculty as well as continuing education to already established staff.

When using a wiki in the classroom, teachers have to be willing to give up control and not use the wiki as an alternative means of distributing material. Student interactions can be encouraged by assigning small teams to create content on the wiki as well as to create individualized pages. Setting social norms and practices for participation can enhance negotiations and knowledge creation. Creating seed postings for students to build on and allowing some disruption of instructor-centered learning activities to pursue student-centered ones can help a wiki flourish.

To date there are no high profile wikis dealing exclusively with biotechnology, though there are more general life sciences sites such as *mybio.net*. The ability to empower students to search, update, and collaborate on wikis provides an incentive for their integration into distance learning programs which has not yet been fully grasped.

VIRTUAL AND REMOTE LABS

For distance science programs one of the outstanding challenges is how to impart lab skills. For traditional distance education programs, such as the infectious disease program at the University of London, this knowledge is passed on through the use of graphics, micrographs, and detailed descriptions. With the interactive nature of the Internet, a number of universities have begun experimenting with virtual and remote labs.

A virtual lab can take many forms, the most simple being a simple pictorial walk-through of procedures, to a more rich multimedia presentation, and up to an interactive simulated environment. A remote lab is a complicated incarnation, being a fully-interactive remote lab controlled online. A remote lab requires much more expense to set up and maintain compared to a virtual lab, but has much more flexibility and real-time feel and enables students to experiment with things which can go wrong, rather than following a scripted, closed environment.

Multimedia presentations with limited interactivity and embedded quizzes have become a staple of many online lab courses. These presentations are often based on platforms such as Adobe Flash which can be easily integrated for use with virtual learning systems and web browsers. There are many examples of virtual labs available on the Internet, with their level of sophistication increasing. However, a 2007 paper in the Journal of Online Learning and Teaching noted that more 85 percent of students perceived a face-to-face lab as more effective than a virtual lab, with large gaps in understanding hypothesis generation, experimental design, and critical evaluation.[18] These results suggest that virtual labs still have a long way to go.

Remote labs, where the real-time control of instruments is available to students through the Internet, are viewed by many distance educators as being the ultimate solution to teaching lab skills online. Initiatives such as MIT's iLabs, funded by Microsoft, have already developed infrastructure for Internet-accessible labs and integrated them into the classroom. However they acknowledge there are many challenges ahead and that the instrumentation involved only gives access to a limited subset of the skills used in a lab environment.

One hybrid option explored at a number of institutions, including UMUC, has been the use of kits at home to conduct experiments complemented by online instruction and guidance. This approach can address some of the criticisms leveled at virtual labs, by providing the hands-on feel and smell of the experiments and concerns by assessment bodies. An alternative approach to this hybrid method is to arrange with a local facility for students to carry out experiments there. It is worth noting that these courses may be the only opportunity for some students to experience the lab environment and have a deeper understanding of the science behind the conceptual ideas taught in the classroom.

---

18 StuckeyMickell, T. A., & StuckeyDanner, B. D. (2007). Virtual labs in the online biology course: Student perceptions of effectiveness and usability. Journal of Online Learning and Teaching, 3(2), 105-111.

CONCLUSIONS

The promise of distance learning and, in particular, online learning, has not yet been fully realized. We are still going through the transition from limited media, classroom-based courses to self-empowerment of students in a distributed, multimedia, virtual learning environment while teaching theory is slowly catching up with technical innovations and implications. The diversity and richness of the Internet along with new tools and resources has the potential to move distance education beyond its current limitations.

The successful integration of current virtual learning environments into educational institutes and the demand for online education is driving the development of expectations and standards. When institutional, faculty, and student concerns are addressed and lessons learned from past failures, everyone can benefit from these advances. With two-thirds of institutions now offering some form of computer-enhanced learning, and a consistent increase in student numbers, some of the historical concerns about online education are being eroded and the benefits realized.

For the technically demanding field of biotechnology, where higher salaries are matched by a demand for more educational qualifications, online learning provides a powerful tool for training this workforce. The limited selection of online distance education programs in biotechnology compared to on-campus courses reflects some of the problems with moving science and technology classes online. Mitigation of these problems through program structure, course design, and working closely with the biotechnology industry can produce very successful online programs.

Moorpark College Biotechnology Program in Industrial Manufacturing

Marie Panec

Marie Panec, CPhil., EdD., is Chair, Life Sciences, at Moorpark College. Marie can be contacted at *mpanec@vcccd.edu.*

Moorpark College is a California community college with two major biotechnology companies in its service area. Baxter Healthcare Corporation and Amgen, Inc. In the mid-1990s both companies needed to develop a workforce of skilled workers in biomanufacturing operations. Baxter Healthcare approached Moorpark College with the idea of establishing a biotechnology training program at the college. Dr. Maureen Harrigan, a molecular biologist, was hired by the college to spearhead the effort. A collaboration was established between operations supervisors from the two biotechnology companies and faculty from Moorpark College. Their goal was to create a college program that in the course of two years would provide students with core knowledge in the sciences and train students in the basic skills required for industrial biomanufacturing operations. There was to be sufficient breadth in the skills training such that an individual could be hired for any entry level position in a biomanufacturing facility and quickly and successfully learn the more advanced skills required for their job.

PROGRAM DEVELOPMENT

After a year of meetings the core classes for the biotech program were agreed to be: *General Chemistry, General Biology, Human Physiology, Microbiology, Statistics,* and basic *Word Processing.* In the fall term of the second year of the program, students would take a one-semester course, *Introduction to Biotechnology*, in which they would learn basic techniques in molecular biology. In the spring term of the second year students would take a series of eight two-week modules. Each class would meet for 40 hours during the two week period with each module covering a particular process in a typical biomanufacturing facility. Students would learn the skills associated with that aspect of industrial biomanufacturing. The modules were: *Plant Design*; *Process Support*; *Cell Culture and Microbial Fermentation*; *Recovery and Purification*; *Formulation, Fill and Packaging*; *Environmental Control*; *QC: Applied Analytical Methods*; and *Validation*. The purpose of offering two-week modules was to allow employees from industry to take an intensive course in a specific skill that they needed for their job or for a job they desired to move into.

Following development of the curriculum, the next two issues to be resolved were identification of a site at which to conduct the biotechnology classes and procurement of industrial biomanu-facturing equipment on which to train students. California State University, Channel Islands (CSUCI) had just received their char-ter from the State of California and was in the initial planning stages of opening its campus to students. The then-vice president offered Moorpark College temporary space in the former hospital on CSUCI grounds in which the biotechnology classes could be held. Amgen generously gave one of their employees time to assist college faculty and staff in setting up equipment and preparing the space for student use. The objective in developing the space was to follow the path a product would take through a biomanufactur-ing facility. A gowning room was set up at the entry of the lab. This opened to a corridor off of which was a dish-washing room and glass storage area. From there the corridor led to a large open lab area which contained fermentation equipment, an autoclave, centrifuges, incubators, and refrigerators. There was counterspace for equipment required for DNA and protein analysis procedures. Off in a separate area of the lab space was a room with biosafety cabinets for tissue culture.

Both Baxtercare Healthcare and Amgen were generous in their

donation of equipment to the fledgling program. Additionally, a National Science Foundation – Advanced Technological Education grant funded necessary supplies and equipment that could not be procured through donations.

The final problem to be resolved before the initial classes could be offered was the issue of who would teach the biomanufacturing processes to the students. The college faculty did not have knowledge or training to teach industrial biomanufacturing. Both Baxter Healthcare and Amgen stepped forward and allowed their employees time to teach biomanufacturing techniques and skills to the students. The teaching of the biomanufacturing classes by industry experts has become a hallmark of Moorpark College's biotechnology program. As a side benefit for industry, this practice has allowed them the opportunity to recruit new employees with known skill sets from amongst the students.

EARLY YEARS OF THE PROGRAM

The first biotechnology classes were offered in Fall 1999. The initial cohort of students was small. Students fell into one of three categories: students seeking their AA degree, students with bachelor's or higher degrees seeking skills training to obtain employment in industry, and students from industry seeking additional training for job promotion at their place of work. Most of the students fell into the second category, already having a four-year degree. This continues to be the most common category of student in our program.

A problem that we have struggled with since the inception of the program and that continues to be an issue is recruitment of students. We work with high schools visiting their classes and career fairs, we do outreach with our own college students advising them of the opportunities in biotechnology, and we attend job fairs at local biotech companies. We have never seen a significant increase in student enrollment as a result of any of these efforts. Most students continue to come to us with degrees and seek job training in the field to make themselves employable. We have yet to determine an effective way to advertise to this population of students. As a result, our class size has remained small.

For the first couple years of the program classes were offered during the day. In the third year after the start of the program we attempted to attract students working in industry by offering the biotechnology classes in the evening. However, this strategy did not result in increased enrollment of students from local industry. We were unable to determine whether this low enrollment was due to an inability to reach those students potentially interested in the program or because industry employees were not interested in the program. What was certain was that regardless of when the capstone biotech class was offered, 20 hours per week of class time was very intense for any student.

After four years with no students from industry interested in enrolling in any of the two-week modules, we decided to make the eight separate modules into a single course covering the same material. Administratively it was much easier, and there was the advantage that if an industry speaker could not make his or her assigned class period, we could juggle the class schedule and still be able to accommodate that speaker at a later date.

PROGRAM CHANGES

In 2003 CSUCI notified us that they needed for their own students the lab space they had been renting to the Moorpark College biotechnology program. We vacated the CSUCI space and moved the equipment and the program onto the campus at Moorpark College. Unfortunately, there is no dedicated space on the Moorpark campus for a biotechnology lab, so conducting the labs proves challenging. Much of the equipment is in storage until we can build a facility to house it. We have found space to set up a temporary cell culture lab and for the remaining labwork we share space with the biology program.

Student enrollment in the biotechnology program reached a critically low point in Fall 2005. The fall class *Introduction to Biotechnology* was cancelled due to insufficient enrollment. This class and the *Industrial Manufacturing* skills class were taught in the Spring of 2006 to allow a final cohort of students to complete the program. The college vice president agreed to continue support of the program if a thorough revamping was done to make the program academically more accessible to students.

Prof. Mary Rees took the academic year 2006-2007 to develop a biotechnology curriculum that would allow students with less of a science background to enroll in the program. Students would take biotechnology classes starting the first semester of the two-year program. The intention was both to encourage students to persist in their studies and to provide them with some employable skills early on. Many students drop in and out of college, taking classes as they have time in their life. If a student "stopped-out," he or she would have learned some basic biotechnology lab skills that they could potentially use in a job situation. Spreading the coursework out over four terms and reducing the number of hours per week also better accommodated the non-traditional student, who could afford only a few hours of coursework each week. We also hoped to attract more industry employees through a career laddering process in which employees could expand their skill sets by taking classes focused on a particular aspect of the industry with the goal of moving up in their career. Prof. Rees met with an industry advisory board multiple times during the curriculum development phase to vet the program with industry representatives and to ensure that the program we developed would meet local industry needs.

To support the development of the biotechnology program more effectively, college administration allocated funding for a full-time tenure track faculty member in biotechnology. In June 2007, we hired Dr. Subhash Karkare, who brings extensive industry experience to the position. He will oversee the future direction of the biotechnology program.

The academic year 2007-2008 is the first year that the revamped program has been offered. There are three significant changes in the skills classes in the revamped curriculum:

1. Pre-requisites for the skills classes have been eliminated. Students can take any of the biotechnology skills class at any time in their educational program.
2. Rather than offering the technical skills in the order that they are used in the development of a product in a biomanufacturing facility, skills now have been clustered together with like-skills taught in a single class.
3. The 20 hour per week skills class has been broken down into five skills classes, each class a full semester. The emphasis is on skills acquisition. The classes are one hour of lecture per week and three hours of lab per week (except for the cell culture class which is 1 and 5).

The hallmark of the Moorpark College biotechnology program, skills classes taught by industry experts, has been retained in the revamped program. Industry specialists in a given area are recruited to teach their particular specialization as a part of the course. The biotechnology skills classes are taught at night, for the purpose of attracting industry employees to the classes.

CURRENT PROGRAM

The program in its new configuration consists of three possible tracks for students to follow depending on their career goals.

- Two-year A.S. Degree in Biotechnology
- Two-year Certificate in Biotechnology
- One-year Manufacturing Operator's Certificate

To receive the A.S. degree students must complete the college's general education transfer requirements. Given all the science courses required for the biotechnology program, this is a rigorous program of study that necessitates the student taking 18

– 20 units every semester for two years in order to complete the program in two years. Students desiring a B.S. and intending to transfer to a four-year institution would take this track.

The Certificate in Biotechnology requires the same core curriculum and science coursework as the A.S. degree, but does not the general education courses. Students seeking employment after completing the program would take this track.

The recommended pattern for the Certificate in Biotechnology is:

- First semester – *English Composition, General Chemistry A, Introduction to Biotechnology and Molecular Biology I,* and *Environmental Control and Process Support* (16 units)
- Second semester – *General Chemistry B, General Biology A, Introduction to Biotechnology and Molecular Biology II,* and *Manufacturing: Quality Control and Validation* (17 units)
- Third semester – *Statistics, Human Physiology* (or *General Biology B*), and *Cell Culture and Microbial Fermentation* (13 units)
- Fourth semester – *Microbiology, Bioprocessing: Recovery & Purification,* and *Business and Governmental Regulation* (9 units)

The one-year program culminating in a manufacturing operators' certification introduces the student to biotechnology, gives the student a basic foundation in biology and chemistry, and teaches the student some basic lab skills that would be applicable in a number of lab-based industries. Students who need to stop-out would take this track.

The recommended pattern for the Manufacturing Operator's Certificate is:

- First semester – *General Chemistry A* and *Environmental Control and Process Support* (10 units)

- Second semester – *General Chemistry B, General Biology A*, and *Manufacturing: Quality Control and Validation* (15 units)

Course descriptions for each of the biotechnology classes are:
BIOT M01A - Introduction to Biotechnology and Molecular Biology I
Examines the role of molecular biology in the manufacturing of commercial pharmaceutical and agricultural products. Introduces basic biotechnology laboratory skills, including documentation, safety, and solution and buffer preparation. Develops proficiency in aseptic technique, spectrophotometry, and electrophoresis (4 sem. units).

BIOT M01B - Introduction to Biotechnology & Molecular Biology II
Expands concepts and techniques introduced in *BIOT M01A*. Introduces modern molecular biology techniques, including nucleic acid isolation, recombinant DNA techniques, cell transformation, recombinant DNA analysis, nucleic acid hybridization, and DNA sequence analysis. Explores the production and purification of recombinant proteins using biochemical techniques such as immunochemistry and chromatography (4 sem. units).

BIOT M02A - Environmental Control and Process Support
Provides skills training in industrial biotechnology with emphasis on manufacturing of biopharmaceuticals. Presents an overview of the manufacturing process and introduces environmental control and process support with a focus on Good Laboratory Practices (GLP) / Good Manufacturing Practices (GMP), clean room procedure, monitoring techniques, and required documentation (2 sem. units).

BIOT M02B Manufacturing: Quality Control and Validation
Provides skills training in industrial biotechnology with emphasis on manufacturing of biopharmaceuticals. Introduces validation and quality control. Reviews manufacturing process, including

formulation, lyophilization, packaging and filling. Focuses on validation assays, systems evaluations, process testing and reporting (2 sem. units).

BIOT M02C - Cell Culture and Microbial Fermentation

Provides skills training in industrial biotechnology with emphasis on manufacturing biopharmaceuticals. Introduces cell culture and microbial fermentation. Focuses on bacterial techniques, microbial assessment, mammalian cell culture, bioreactor fermentation, and media preparation. Compares small and large-scale cell culture (3 sem. units).

BIOT M02D - Bioprocessing: Recovery and Purification

Provides skills training in industrial biotechnology with emphasis on manufacturing biopharmaceuticals. Introduces bioprocessing, recovery, and purification techniques. Focuses on protein separation and purification, chromatography, large-scale recovery, and identification assays. Reviews skills necessary for a successful job search in the field of biotechnology (2 sem. units).

BIOT M02E - Business Practices and Governmental Regulation

Provides skills training in industrial biotechnology with emphasis on manufacturing biopharmaceuticals. Examines manufacturing from the perspective of company operations, including general building design, logistics, and bioethics. Focuses on business practices and governmental regulations (2 sem. units).

STUDENT RECRUITMENT

Student recruitment presents the most significant challenge for the future of the biotechnology program. We have an excellent program in a growing industry. However, many students do not have an understanding of the career possiblities in biotechnology. Nor do many high school students who have the science background to succeed in the field have an interest in a two-year technical program. The students we are most likely to continue to attract are individuals with four-year degrees who are looking for skills acquisition to enter the industry. Identifying these individuals and promoting our program to them will continue to be a challenge. Additionally, we would like to attract more of non-degreed and beginning college students to the program. Most of our recruitment efforts to date have focused on promoting the program to high school and beginning college students and informing them about the opportunities in the field of biotechnology.

Recruitment efforts have included:
- A summer bridge program for high school students to introduce them to the world of biotechnology. This 5-day, 25-hour program emphasizes hands-on exercises that catch students' imaginations. We have institutionalized this program as a course offered in the spring and/or summer.
- Tours of the biotechnology lab space for high school counselors along with an opportunity for them to ask questions and providing them with program brochures.
- Visits to local high school classes and career days to demonstrate a biotechnology lab exercise and discuss with students the opportunities available in the biotechnology industry.
- Visits to local industry job fairs to provide information about our program to industry employees interested in upgrading their skills.
- Publicity in college outreach brochures and schedule of classes.
- Information booths staffed with faculty at the college Parents' Evening and academic fairs.
- Visits to college general science classes to inform college students about the opportunities in the field of biotechnology.
- Information sent to college counselors about the program and the classes we are offering.

FUTURE PLANS

Moorpark College has approved funding for a new health sciences building, the second floor of which will be devoted to the biological sciences. One lab and several associated side rooms have been allocated for the biotechnology program. We anticipate breaking ground on the building in Summer 2008 with a completion date scheduled for Fall 2010.

Johns Hopkins Biotechnology Education: An Integrated Program of Multiple Degrees, Concentrations, and Collaborations
Lynn Johnson Langer

Lynn Johnson Langer is Senior Associate Program Chair in the Advanced Biotechnology Studies Program at Johns Hopkins University. Lynn can contacted at *ljlanger@jhu.edu*.

Biotechnology continues to expand rapidly with new discoveries, creating life-saving products at a rapid pace. The industry, by definition, is a merger of science and business. This demands a multi-disciplinary workforce skilled in basic science research, product development, regulatory affairs and commercialization. Johns Hopkins University offers students the opportunity to study with industry leaders and integrate meaningful, practical coursework, with their own needs as working professionals.

Biotechnology in the United States began on its current course in the mid 1980s as scientists at the National Institutes of Health (NIH), and elsewhere left to form their own companies. Maryland, the home of the NIH, has been an integral part of this growth. Within ten years, Maryland had the nation's third largest concentration of biotechnology companies, in part because of the huge brain trust located in Maryland from the NIH, the FDA, and Johns Hopkins University. In the early 1990s the Dean of the School of Arts and Sciences, Lloyd Armstrong, tasked the school with investigating the possibility of establishing a part-time Master's Program in Biotechnology in response to the growing biotechnology industry in Maryland. The Master of Science in Biotechnology was developed and became the first of four master's degree programs in what is now Advanced Biotechnology Studies

(ABS).

CONNECTION WITH INDUSTRY-RELEVANT ISSUES

The primary goal of the ABS program is to provide a relevant curriculum taught by the qualified professionals recruited from academia, industry, and government agencies and laboratories. The curriculum provides a solid foundation in the life sciences, which is the basis for the biotechnology field, and integrates the ideas, tools, and principles of business that are required to develop marketable products for healthcare, protecting the environment, and enhancing agricultural production.

Quality of the program has been a major consideration. The program is administered by the Biology Department within the Krieger School of Arts and Sciences with an academic program chair—a tenured professor from the biology department—who oversees the program. Currently, there are three Associate Program Chairs (APCs) who are faculty/administrators, and who teach in the program and run the daily activities of the program. The APCs serve in their area of expertise, including bioscience, bioinformatics, regulatory affairs, and the business of biotechnology while connecting with the students as their advisor and instructor. To ensure program quality, the APCs and the chair review all new programs with the academic council and the dean.

The APCs actively participate in the biotechnology industry and regularly attend and speak at relevant conferences. This participation as a full partner in the biotechnology community and industry allows the program to be at the forefront of new ideas and initiatives. For instance, bioinformatics and bioscience regulatory affairs are relatively new disciplines with few formal training opportunities available for practitioners in the field. Through formal and informal conversations with industry leaders, faculty, and students, Johns Hopkins recognized the need for more formal education in these fields.

Located in the heart of one of the nation's leading regional biotechnology clusters, ABS offers students the ability to learn, advance and succeed in this field with a variety of learning opportunities designed to meet the needs of working adults. The Master of Science in Biotechnology was the first degree program in ABS and is grounded in biochemistry, molecular biology, and cellular biology. However, ABS understood early in the development of the program that although biotechnology is based on science, it is an applied discipline that must involve business and regulatory aspects. To integrate these disciplines, courses developed for the program included *Economic Aspects of Biotechnology* and *Legal Aspects of Biotechnology.* Today, the MS in Biotechnology includes concentrations in Enterprise, Regulatory Affairs, Bioinformatics, Biodefense, and Molecular Targets and Drug Discovery.

By 2005, the program grew from a single master's degree to five degrees. ABS now offers the Master of Science in Bioscience Regulatory Affairs; the Master of Science in Bioinformatics, offered jointly with the Whiting School of Engineering; the MS in Biotechnology/MBA offered jointly with the Carey Business School; the Certificate in Biotechnology Enterprise; and a fellowship in Molecular Targets and Drug Discovery with the National Cancer Institute. The programs and degrees are geared toward full-time working adults, so courses are offered on weekends, evenings, and online.

ON-SITE AND DISTANCE EDUCATION

ABS offers over fifty courses online. The MS in Biotechnology, the MS in Bioscience Regulatory Affairs, the MS in Bioinformatics and the Certificate in Biotechnology Enterprise may all be completed fully online. Courses are typically taught for two semesters on-site before being developed into online courses. The course content and materials are the same for on-site and online classes; only the course delivery systems are different. The online courses are highly interactive and students engage regularly with their instructors and their classmates. In fact, data collected from student evaluations indicate there is no difference in satisfaction for on-site or online courses.

MASTER OF SCIENCE IN BIOTECHNOLOGY

The Master of Science in Biotechnology is the cornerstone of ABS. The curriculum is designed so that graduates can engage in research, lead lab teams, help to make development and planning decisions, and create and apply research modalities to schemes set in large research projects. Also, because many students enter the program as managers, marketers, lawyers, and liaisons, these students are able to effectively bridge the worlds between their roles and the scientists in their organizations. This allows for effective communications and decision making.

Students entering the program must have an undergraduate degree in the natural sciences or engineering with at least a 3.0 grade point average, two semesters of college-level biology and two semesters of organic chemistry. Students are required to take a total of ten courses: four core science courses and six electives. Most students choose not to concentrate in a specific area, but to customize their electives to suit their specific purposes. Students may, however, choose to concentrate in biodefense studies, biotechnology enterprise, bioinformatics, regulatory affairs, and molecular targets and drug discovery technologies.

CONCENTRATIONS

BIODEFENSE

The biodefense concentration integrates basic and translational science to train the next generation of professionals for employment in academia, industry, and government. The curriculum provides students with a solid foundation in basic science and investigates the various applications of medical science and biotechnology for detection, identification, and response to biothreats. Specific disciplines of study include molecular biology, infectious diseases, bioinformatics, immunology, epidemiology, molecular diagnostics, and policy.

BIOTECHNOLOGY ENTERPRISE

For research discoveries to reach the public, an understanding of the overall enterprise of biotechnology is essential. Success requires two distinct sets of skills and perspectives: understanding the science and understanding the business. Students in this concentration must complete four core science courses, four core enterprise courses, and two science electives.

BIOINFORMATICS

Given the vast amounts of information generated from studies on humans and other organisms, and the need of scientists and researchers to access and manipulate these data, the biotechnology program offers courses that can either be sampled individually or taken together to make up a concentration in bioinformatics.

REGULATORY AFFAIRS

The Regulatory Affairs curriculum was developed in consultation with representatives from the Food and Drug Administration (FDA), the Regulatory Affairs Professional Society (RAPS), and the biotechnology industry, this concentration provides students with the knowledge required for organizations to comply with federal and state regulatory statutes for the development, approval, and commercialization of drugs, biologics, foods, and medical devices. Students in this concentration must complete four core science courses, four core regulatory affairs courses, and two electives.

MOLECULAR TARGETS AND DRUG DISCOVERY TECHNOLOGIES

This concentration was developed with the Center for Cancer Research/National Cancer Institute (CCR/NCI)[1] and integrates the in-class didactic training and hands-on laboratory experience required for graduates to contribute to the advancement of knowledge and research in the field of drug discovery.

---

1   http://ccr.ncifcrf.gov/careers/jhu

NATIONAL CANCER INSTITUTE MOLECULAR TARGET AND DRUG DISCOVERY FELLOWSHIP

Johns Hopkins University and the Center for Cancer Research/ National Cancer Institute (CCR/NCI) have developed an innovative graduate program that prepares the next generation of scientists in drug discovery technologies. Fellows earn a Master of Science in Biotechnology with a concentration in Molecular Targets and Drug Discovery Technologies, participate in important cancer research, work in CCR/NCI laboratories. Students receive paid tuition for up to two years and an annual stipend.

Up to five students per year are selected for the two-year fellowship. Fellows earn a Master of Science in Biotechnology with a concentration in Molecular Targets and Drug Discovery Technologies, participate in full-time cancer research at the National Cancer Institute laboratories in Bethesda or Fredrick, MD, and receive full tuition and a competitive stipend. Fellowships are renewed annually. Admission is highly competitive and candidates must be a recent (within the past 3 years) graduate of an accredited university or college, must be a U.S. citizen or permanent resident, and must be accepted into the Master of Science in Biotechnology program as a degree student before being considered for the fellowship.

CERTIFICATE IN BIOTECHNOLOGY ENTERPRISE

Students who seek a solid understanding of the business aspects of biotechnology to advance in this rapidly changing field, but who are also well versed in the scientific aspects, can apply to the Certificate Program in Biotechnology Enterprise. Students can be admitted to the program without prior graduate level work, but a bachelor's degree in the life sciences is recommended. Students who successfully complete the certificate and subsequently decide to seek admission to the master's degree program in biotechnology will receive credit for five of the courses taken in the certificate.

MASTER OF SCIENCE IN BIOSCIENCE REGULATORY AFFAIRS

The products and services offered by biotechnology companies are generally highly regulated. Companies that develop diagnostics and provide reagents and supplies must provide stringent quality control to ensure FDA good manufacturing practices are followed. Companies report great difficulty in finding trained professionals educated in bioscience regulatory affairs and with the necessary skills to fulfill federal and state regulatory requirements. To meet this need, ABS created the Master of Science in Bioscience Regulatory Affairs, using the expertise of professionals from the federal government, industry, and academia.

An increasing number of students in the MS in Biotechnology program started taking courses related to the regulation of new products. This led to the establishment of an advisory board including a former acting commissioner of the FDA, industry leaders and academics. The board was assembled to determine the type of education needed for workforce development. The advisory board recognized a strong need for higher education in regulation and compliance. The board decided that a full master's degree, rather than a certificate program, would best serve the industry and The Master of Science in Bioscience Regulatory Affairs was created and launched in fall 2005. All the courses in the program were approved and/or created by the advisory board. The first graduates completed the program in May 2007. Many of these graduates have gone on to hold mid-to-senior level positions in the field of regulatory affairs.

The program is grounded in the sciences and all students entering the program must have taken at least one undergraduate course in biochemistry and cell biology. If they have not taken these courses previously, the program offers a pre-requisite course, *Bioscience for Regulatory Affairs*. Students are required to complete ten courses in the program, including seven required courses and three electives. Students take *Biological Processes, Introduction to Regulatory Affairs, Product Development, Introduction to cGMP, Clinical*

Development of Drugs and Biologics and a practicum. The practicum is the final course in the program and students work on teams to solve real-world problems, both from the federal regulating agencies perspective and from the industries perspective.

MASTER OF SCIENCE BIOTECHNOLOGY/MBA JOINT DEGREE PROGRAM

The innovative MS/MBA program allows students to gain deeper understanding of bioscience while developing business skills in areas such as accounting, negotiation, finance, and regulatory and legal matters. Drawing on the strengths of the Zanvyl Krieger School of Arts and Sciences, and the Carey Business School, the program allows students to work side-by-side with expert faculty in biotechnology and business. The program prepares students for success in both the science and business of biotechnology and allows students to earn two advanced degrees in less time than it would take to earn them separately.

Success in biotechnology requires understanding both the science and the business of biotechnology. Unfortunately, most professionals in the industry possess one or the other. As a consequence, scientists and business people in the field often do not "speak the same language." More fundamentally, they have difficulty in critically evaluating the implications of developments from both a scientific and business perspective. Business professioanls are often at a disadvantage in analyzing the scientific potential for a particular advance, and scientists are at a disadvantage in analyzing the business potential of that same advance. This issue creates problems for an industry that needs to manage effectively its inherent technical and financial risks.

The integrated MS Biotechnology/MBA program in biotechnology at Johns Hopkins University has three purposes:
- Impart the knowledge and skills in the principles and science of biotechnology and business that will enable students to be effective managers and leaders

in biotechnology-related agencies and organizations
- Integrate the science and business curricula in a way that makes the necessary connections while respecting the intellectual integrity of the two fields
- Streamline the degree-earning process to maximize the relevance and effectiveness of courses and subject areas

Those who complete the proposed program are able to apply the core principles of science and business to the biotechnology industry, to be fluent in the language of science and business, and to ask the right questions of scientists and businesspeople on critical issues facing their organizations. Additionally students in the program are able to identify, evaluate, and act on (as appropriate) scientific and business opportunities that arise in the field. Graduates from this program are equipped to assume senior responsibility for promoting and leading biotechnology initiatives in their organizations.

MASTER OF SCIENCE IN BIOINFORMATICS

ABS also developed the innovative MS in Bioinformatics program that prepares bioscience and computer science professionals for success in bioinformatics. Drawing from the strengths of the Zanvyl Krieger School of Arts and Sciences and the Whiting School of Engineering, this program fully integrates the computer science and the bioscience needed to excel in the dynamic new field of bioinformatics. Students entering the program must have a strong foundation in biochemistry, programming using Java, C++ or C data structures, statistics, and calculus.

Students are required to take eleven total courses: five core courses including, *molecular biology, gene organization and expression, foundations of algorithms, database systems,* and *computers in molecular biology.* Students then choose from four concentrations courses and one elective from the biotechnology courses and one elective from computer science courses. After completion of the core and concentration courses, students may choose an independent study project as one or both of their electives. Students have up to five years to complete the program and may do so in the classroom, fully online, or as a combination of on-site and online classes.

STUDENTS EDUCATED FOR INDUSTRY NEEDS

The APCs work as leaders and committee members throughout the industry. As such, they interact regularly with industry leaders to learn what areas are of current interest. The program faculty are practitioners in the field and come from major companies and research institutes such as the National Institutes of Health, the Food and Drug Administration, the US Army Medical Research Institute of Infectious Diseases, MedImmune, Merck, and Human Genome Sciences, to name a few. The APCs hold memberships on the Maryland Governor's Workforce Investment Board, the Regulatory Affairs Professional Society, the American Society of Microbiology, and the Biotechnology Research Institute. The program collaborates with federal agencies such as the National Cancer Institute, the Food and Drug Administration and other universities such as Peking University in Beijing, China.

METRICS FOR SUCCESS

Students fill out evaluations after every course, where they have an opportunity to evaluate the course and the instructor, and to provide ideas for other courses. The evaluations are closely monitored and are used by faculty to strengthen existing courses and to create new courses. Information is also gathered from students when they graduate that describes their overall experience within the program and ideas they may have for improvement.

ABS is unique in the biotechnology industry in that it offers education that covers all biotechnology segments. Faculty are real-world experts who come from academia, industry, and government agencies. ABS is situated in the heart of the Maryland Biotechnology industry, located near the National Institutes of

Health, the Food and Drug Administration, and major defense laboratories. The program attracts outstanding faculty who find teaching in their areas of expertise a wonderful opportunity. Additionally, the online courses offer the wealth of faculty knowledge to students around the world. Thousands of students have graduated from the ABS, and graduates of the programs now run biotechnology firms, hold senior executive positions in industry, and run research laboratories. Students hail from around the world, and faculty are also located internationally. While the program is global, and collaborations between the university, multinational companies, and international governments are extensive, because biotechnology is a cross-border discipline, students continue to find a learning community in which they belong.

The University of Toronto Master of Biotechnology Program: A New Model for Graduate Education in Biotechnology

Leigh Revers and R. Scott Prosser

Leigh Revers, MA, DPhil, is Assistant Director of the Masters of Biotechnology Program at the University of Toronto Missisauga. Leigh can be contacted at *leigh. revers@utoronto.ca.*

R. Scott Prosser, Ph.D., is Associate Professor and Director of the Masters of Biotechnology Program at the University of Toronto Missisauga. Scott can be contacted at *scott.prosser@utoronto.ca.*

Educators are painfully aware of the mismatch between conventional graduate education in scientific disciplines and the needs of technology-intensive industries. A 2003 survey of 1992-1998 doctoral graduates from top-10 National Research Council-ranked chemistry departments revealed that a significant fraction of respondents felt that their education was deficient in training, career preparation, and supervisor mentorship.[1] In a separate study involving graduate students from eleven arts and sciences disciplines, less than half (43.7 percent) of the respondents believed that their training had adequately prepared them to collaborate in interdisciplinary research.[2] Although the current model of thesis-based graduate education may serve the institution well in the sense that the research interests of the supervisor

---

1   Kuck, V. J.; Marzabadi, C. H.; Buckner, J. P. & Nolan, S. A. (2007) *A review and study on graduate training and academic hiring of chemists. Journal of Chemical Education 84(2):277-284.*

2   Golde, C. M. D., T. M., *The Survey of Doctoral Education and Career Preparation: The Importance of Disciplinary Contexts. In Path to the Professoriate: Strategies for Enriching the Preparation of Future Faculty (Donald H. Wulff, Ed.) Jossey-Bass: San Francisco, 2004.*

are satisfied, as educators we must also ask how the needs of the graduate student might be better served if his or her intention is to enter the workforce in a science-intensive business.

Today's existing doctoral and post-doctoral career paths are designed to train graduate students to be successful and independent researchers. However, these students often find themselves with an overly-focused research background, little or no experience in an industrial setting, no formal training in management, and minimal opportunities to develop their interpersonal skills. Only a modest percentage of students graduating from top-tier North American universities, and a very small percentage of students graduating from middle-tier establishments, secure academic positions in universities or colleges. With proper training, many of the remaining students would be far better placed to find fulfilling careers in a technology-intensive industry, where their scientific training could have real impact. Our experience has been that in addition to research and development opportunities, students with a strong scientific background and an understanding of corporate culture can quickly acquire the skills necessary for sales and marketing, management, product development, manufacturing, and regulatory affairs. In this chapter, we outline a new two-year professional Master's program at the University of Toronto, developed in response to the needs and expectations of industry in the pharmaceutical and biotechnology sectors. Our intention was to address some of the shortcomings of traditional graduate programs in science, and to create a program tailored to students expressing a desire to apply their scientific training in an industrial context in the biotechnology sector. The program emphasizes the continual development of interpersonal, problem-solving and management skills, job placements in at least two work environments, and a regimen of graduate-level laboratory and classroom science and business courses.

THE LOCAL BIOTECH AND PHARMA ENVIRONMENT

"Cash is king" is an ubiquitous dictum of the entrepreneur, and perhaps nowhere is it more relevant than in the biotechnology sector. New biotechnology ventures are frequently cash-poor, surviving on a life-support system of carefully tranched, and usually under-capitalized, investments from venture capitalists that are carefully meted out in reward for achieving what is almost as commonly, a very technically demanding series of milestones. The technologies espoused by such companies are most likely proof-of-concept debutantes, offering a potential service or product that requires extensive development before it can realize any tangible value for its investors. The key word here is *potential*. As such, small biotechnology companies do not qualify for conventional business loans, instead settling for equity financing, possibly laced with the odd debenture or two, and, if they are lucky, a dash of grant money that the scientific team has fought for successfully from one or other of the governmental agencies. Indeed, in November 2006, Gary Pisano wrote tellingly[3] in the Harvard Business Review that the current business model for biotech industry was, in essence, broken. No wonder such companies, with their expensive laboratories to run, large capital outlays on state-of-the-art equipment, and the inescapable daily consumption of high-cost, single-use disposables, as well as a daunting succession of regulatory hurdles to clear, face an improbable and arduous path to climb before they can become profitable.

While the foregoing is true throughout North America, in Canada, and particularly in Ontario, the situation is even less appealing, and many young ventures are understandably flagging. Prior to the millennium, the future for biotechnology in Ontario looked bright;[4] and until recently, 40 percent of the value of new

---

3   Pisano, G. P., (2006) Can science be a business? Lessons from biotech. *Harvard Business Review* 84(10):114-125.

4   McAuley Endersby, J. (1999) Kick-starting biotechnology in Ontario. *Nature Biotechnology* 17, 444-446.

Canadian venture placements were funded by Ontario, owing largely to the existence of the province's labour-sponsored investment funds (LSIFs), which were originally intended by the Ontario government to stimulate research and development, and to create high quality jobs.[5] Under this scheme, investors were able to choose from among almost 100 LSIFs, which invest in small companies, and receive substantial tax benefits. Then, on September 30 2005, the Ontario Minister of Finance announced that the LSIF tax credit would be eliminated by the end of the 2010 tax year, and, unsurprisingly, the last two years have seen a major decline in small venture investments.

The ramifications for young firms were palpable, with overall venture capital investment across Canada dropping in 2006 to 75 percent of its 2005 figure, and investment in early-stage companies in Ontario falling to a mere 10 percent of its 2000 level.[5] The biotechnology industry, with its temperamental reputation as a high cost, high risk-reward, sector has been especially hard-hit, with the year-on-year dollar figure of annual investments projected to drop by 48 percent in 2007.[6] In a survey this year by BIOTECanada, one of the federal government agencies monitoring the biotechnology sector, and PriceWaterhouseCoopers,[7] CEOs were asked to list the top three most challenging hurdles they felt their companies would face over the next two years. Not surprisingly, the most pressing issue cited by respondents was their organization's ability, or otherwise, to access venture or other types of capital, leading the survey's authors to identify this as the recurring theme across the industry.

Despite this recent and seemingly ubiquitous litany of doom and gloom on the financial front, government officials and industry gurus are nonetheless remarkably buoyant. Biotechnology and

---

5   Wanless, T. *Tech Startups Find it Tough to Raise Cash. In The Angel Journal*, http://www.theangeljournal.com/cms/content/view/27/114/

6   Corr, T. (2007) *Seminar: Introduction to Technology Commercialization, Entrepreneurship 101*, MaRS Discovery District, Nov 7.

7   *Canadian Life Sciences Industry Forecast 2007*, BIOTECanada & PriceWaterhouseCoopers.

emerging life science companies are widely considered to be in the vanguard of the high-tech economic sectors, alongside other high profit businesses in information technology and telecommunications; and they are collectively touted as the keys to a bright economic future for western developed nations. BIOTECanada's forecast neatly distills the situation by remarking that their findings for 2007 "showcase the industry's maturity; however, the industry still faces significant challenges in…the battle for capital and the need to recruit industry-savvy people." So while the availability of cash remains enthroned as the number one enemy to small biotechnology firms' success, access to an abundant supply of talent is perceived as a close second.

Most commentators in the biotechnology space agree that there is, and will for the foreseeable future continue to be, a growing demand for the highly skilled individual with scientific credentials and a mind for business. Indeed, returning to BIOTECanada's survey, second on those same executives' lists were the problems faced by their companies in attracting and retaining the right kind of employees. In particular, those surveyed expressed concern over the recruitment of personnel of a sufficiently high caliber required for key roles in the multifaceted and challenging processes of product and business development; and they also identified senior management positions, such as Directors of Regulatory Affairs, Chief Executive Officers, VPs of Business Development, Directors of Clinical Development, and VPs of Sales as especially difficult to fill. This, of course, comes as no surprise at all to the cognoscenti, who have long known that the life sciences sector, with its elaborate and demanding regulatory approvals framework, and its *de facto* reliance on cutting-edge developments in the technically complex fields of physical and molecular sciences such as chemistry, molecular biology, pharmacology, immunology, pathology and medicine (all with or without the optional 'molecular' prefix), to name a few, requires a unique repertoire of skills. This is borne out by recent trends among human resources specialists, who are promoting a new vernacular where terms such as 'interdisciplinarianism' and 'interprofessionalism' have become commonplace, and the individuals espousing them are highly prized.

Clearly, it is beyond the scope of this chapter to examine the financial bugbears currently laying siege to the Canadian biotechnology sector, but taking into account the widespread optimism about the economic and societal importance of biotechnology, as educators, we have taken the long-term view that more carefully-tailored learning opportunities must be made available, if Canada is to compete at the forefront of biotechnology commercialization. This credo has presided over the inception of the Master of Biotechnology Program, and in the following sections, we hope to offer insights into the educational strategy we have adopted to supply the life sciences industry with the high quality entry-level individuals that the broad consensus believes are essential for the sector to grow and flourish.

THE MBIOTECH PROGRAM AT THE UNIVERSITY OF TORONTO: AN OVERVIEW

So what, then, is the MBiotech Program? Distilled to a single sentence, it is a 24-month, course-based professional degree at the University of Toronto (U of T) incorporating both science and business courses, coupled with 8-to-12 months of work experience in the biotechnology and biopharmaceutical sectors.

Since its inception in 2001, the program has sought to include leading industry players in molding a professionally orientated academic program that satisfies their needs. Our current board of directors includes academic faculty members, prominent alumni, and key opinion leaders in the biopharmaceutical sector, including representatives from such major pharmaceutical companies such as Amgen, AstraZeneca, GlaxoSmithKline and Merck Frosst. As such we have undergone several iterations in the design and implementation of courses, in a concerted effort to meet the evolving needs of our students, and to maintain a high degree of relevance to the biotechnology and health sciences industry, not only locally, but on a global scale.

The MBiotech Program is also distinctively interdisciplinary.

Our courses are taught by experienced professionals in the biotechnology sector, in addition to U of T teaching faculty drawn from the departments of Biology, Chemistry and Management, as well as additional faculty appointees who are specifically allocated to our program on a full-time basis. As we discuss in the following sections, our curriculum runs the whole gamut, from contemporary bench science, through pre-clinical and clinical drug development, to management fundamentals and organizational skills, as well as offering foci on technology innovation and entrepreneurship. Students then have the chance to put their learnings into action in the real-life context of one or more organizations as part of our highly successful cooperative work-term scheme. We believe that it is this combination of continual training in interpersonal and verbal skills, extensive classroom and laboratory training, and advanced job placements in industry, plus a strong focus on teamwork that provides our graduates with a truly interdisciplinary and professionally relevant educational experience.

SCIENCE, BUSINESS AND THE CO-OP WORK EXPERIENCE: THE HOLY TRINITY OF MBIOTECH

THE MBIOTECH ACADEMIC COURSES

It has been our aim from the inception of the MBiotech Program to provide an intensive series of lecture courses in the first year of the program that provide a solid foundation in the biotechnological space. These courses encompass both the scientific and business realms, with overlapping content that is designed to reinforce key sector-specific knowledge sets, and to provide students with a broad interlocking mosaic of concepts. These academic courses fall into three categories, which we elaborate below. Figure 1 summarises how the different courses are structured over the program's two year duration.

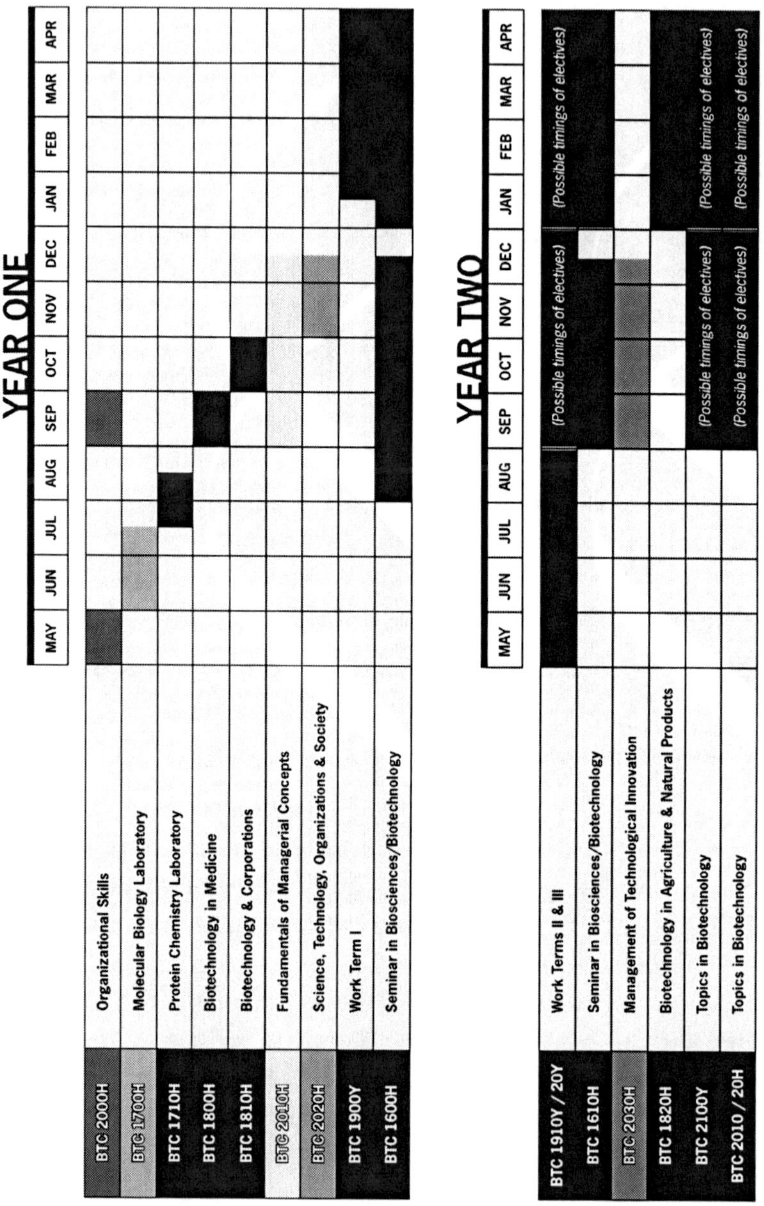

Figure 1: MBiotech courses

I.   TWO PRACTICAL SCIENCE COURSES

It is our firm belief that students at the master's level require a strong grasp of the practical realities of laboratory science in order to progress in the biotechnology industry. While most students go on to accept work-term placements in the larger pharmaceutical companies, such as GlaxoSmithKline and AstraZeneca, and may never engage in bench-work of the type ubiquitous among thesis-based master's programs, the laboratory is, in our opinion, an essential learning ground where future managers and cross-pollinators can hone experimental skills that are, and are likely to remain, mainstays of the industry. Such knowledge and experience, while valuable in and of itself, serves to enhance the communication channels and reinforce respect between managers and scientific staff for whom the techniques and methods of laboratory science are the focus of their daily activities. In essence, master's students emerging from the MBiotech Program who aspire to middle and senior management will be well equipped to understand the challenges faced by technical staff by drawing upon their own first-hand experiences, and, therefore, to manage better.

The MBiotech Program offers two mandatory courses in this realm. The first of these, *Molecular Biology Laboratory (BTC 1700H)*, is delivered annually in May, and, alongside *Organizational Skills (BTC 2000H*, see below), marks the inauguration for students entering the program in their first year. The aim of the course is to introduce and expand upon fundamental experimental techniques commonly used in biomedical research. It is intended to provide hands-on experience working with nucleic acids and proteins, and comprises an intensive six-week schedule, running on weekdays throughout May and June. During the first, introductory week, students are provided with an overview of key protocols from a practical perspective via a short semi-formal lecture series, and are provided with same-day, interactive demonstrations of the corresponding experimental techniques in a fully equipped 'wet' laboratory. In this phase of the course, students group together in teams for the entire duration of the course and familiarize themselves with standard laboratory methods and routines, and adjust to the team environment. The introductory week is followed by an extended research assignment in which the student teams collectively work towards expressing and isolating a biomedically relevant, recombinant protein. Teams are effectively galvanized by this tightly scheduled project, and must successfully design and implement an appropriate research strategy, conduct and manage experiments (based on the standard operating procedures provided), collect and analyze data, and prepare and submit their product with appropriate accompanying documentation to meet a pre-arranged deadline. In addition, teams are expected to formulate their ideas about the potential, relevant biomedical applications of their product, examine any pre-existing intellectual property, and submit concise written and oral reports at the end of the course.

Typical experimental techniques covered in the course range from: simple plasmid DNA isolation, polymerase chain reactions, basic molecular cloning techniques, heterologous expression in bacteria or yeast, recombinant protein affinity purification, Western immunoblotting, and functional (e.g., enzymatic) assays or immunoassays.

The second course, *Biotechnology & Chemistry (BTC 1710H)* runs in July and August. This laboratory is intended to familiarize students with materials chemistry as it applies to biotechnology. In particular, biotechnology applications involving polymer chemistry, nanochemistry, protein and small molecule chemistry, and various hybrids of the above are presented to the students in the form of formal lectures by the instructor, as well as reviews of the current literature. Much of the material is covered by student team presentations (as many as two per day). In this way, students learn to survey the literature and hone their verbal and team-working skills at the same time. Laboratory modules, which are also assigned to teams, encompass a range of advanced analytical techniques, including electrochemical biosensors, surface plasmon resonance, nuclear magnetic resonance, and mass spectrometry. Additional laboratory modules focus on peptide synthesis, protein expression and chemical modifications such as PEGylation or fluorescent tagging, in addition to a plethora of hands–on experience with proteins and peptides, including biosynthetic labeling, and various assays associated with protein expression and activity. Finally, several laboratory modules delve into bioinformatics and structural modeling. The students are expected to incorporate this knowledge and develop a science-based proposal that addresses a municipal, regional, or national problem that could potentially be resolved by biotechnology. The business aspects of this proposal are later examined in greater depth in *Biotechnology & Corporations (BTC 1810H*, see below), where students are required to develop a business plan. The workload in *BTC 1710H* is designed to exceed the capacity of any individual. In this way, we feel we can best develop team skills while simultaneously providing students with the requisite knowledge in chemistry as it pertains to biotechnology.

II.   FOUR BUSINESS COURSES

The program kicks off its repertoire of business courses in the early summer of the first year with *Organizational Skills (BTC 2000H)*. This course introduces the incoming students to the basic skills and concepts needed to become an effective member of an organization, and to some of the fundamentals of organizing technical and commercial work. We feel that the positioning of this course at the beginning of the program is essential, since the large majority of students arrive with little or no management experience, and, as scientists, are naïve to some of the elementary concepts of team-play. Consequently, *Organizational Skills* is designed to allow students to explore various structures to learn about issues of co-operation, teamwork, leadership and goal-orientation. Some of the targeted learnings involve developing: (i) basic people skills, understanding interpersonal differences, and concepts of motivation and leadership; (ii) basic communication skills, both written and oral; (iii) basic team-working skills, and understanding team roles and dynamics; and (iv) fundamental organizational skills, and familiarity with different organizational structures and cultures. This course is also especially relevant to the Program in a more global sense, as it is here that we first define and organize the students into teams that will carry forward to complete the subsequent laboratory modules (*BTC 1700H* and *BTC 1710H*, see above). In addition, *Organizational Skills* serves as a platform for the identification of leadership skills among the students, and provides, for many, the first opportunity for a team approach to problem-solving. The course also seeks to refine oral presentation skills, which are a major requirement of subsequent courses, with the assistance of professional instruction and the use of video recordings for personal critiques. While the majority of the teaching in this course (six sessions) is delivered in the summer, a series of four follow-up modules are scheduled for the fall, allowing the instructor to review the students' progress after a mid-year change in the team line-ups, and also to begin to prepare them for their imminent work-term placements, which begin the following January.

In the fall of the first year, students follow *BTC 2000H* with a foundational business course, *Fundamentals of Managerial Concepts (BTC 2010H)*, which introduces a number of the critical, practical aspects required to operate successfully in today's biotechnologically-focused businesses. Topics covered include elementary financial statement analysis, financial management, and marketing management, as well as some aspects of organizational behavior (reinforcing the learnings delivery by *BTC 2000H*) and strategic management. Theory and application are combined through the conventions of set readings, the preparation and class discussion of selected case studies, and the submission of a team-based project. Following and overlapping with *BTC 2010H* is the third business course in the program, *Society, Organizations & Technology (BTC 2020H)*, a course that examines more closely the economic and business environment in which the biotechnology and biopharmaceutical industries operate. Setting these key industry sectors within the broader context of the development of technology-based industries, issues covered by this course include the role of government regulation, business strategy, and intellectual property rights. *BTC 2020H* further examines the growth and structure of the industry, together with the key strategic issues faced by today's firms. Business concepts that concern the processes of innovation are also introduced, including industry analysis, competition policy, licensing, as well as innovations in the realm of financing. Once again, the course relies heavily on group assignments, in which the student teams study either a specific issue concerning the impact of biotechnology or a case study and then discuss the commercial consequences in class via oral presentation or debate.

The final business course in this quartet is *Management of Technological Innovation (BTC 2030H)*, which is delivered in the second year of study, following an extended absence by the students on their first two work terms (see later). It is also the final course in the program prior to graduation. As such, its explicit goal is to put together many of the components developed in earlier courses, to allows students to build their own technology business plan, either for a new venture, or to create a new strategic business unit within a larger organization. *BTC 2030H* focuses on the factors that enable technological innovation, and contemplates how these factors can stimulate the development of new business opportunities. Students acquire an understanding of the different types of business models that can be developed, looking not only at the dichotomy of licensing versus new venture creation, but also at how this can lead to a variety of value propositions and market segmentation opportunities. The course deals especially with the importance of addressing adoption issues, and of developing a strong intellectual property position and strategy. It also reviews some of the key requirements for success for both entrepreneurs and intrapreneurs. Finally, *BTC 2030H* aims to explain how the different strategic options come together in a financial plan, which underscores a new business plan, and leads to the creation of a document that is pitched to internal or external stakeholders. To enhance the value of the course, students are encouraged to work with real ventures: this fosters a more dynamic learning opportunity, which, we feel, delivers a comprehensive conclusion to the business side of the MBiotech Program.

III.   TWO HYBRID COURSES

The third element of the MBiotech academic course structure comprises two novel science-meets-business 'hybrid' courses that have been specially developed by the faculty to address key areas of overlap at the junction of science and business. Owing to geo-economic considerations and the nature of the majority of the work-term placements, both courses deal heavily with the medical sector, although efforts are also made to examine other realms, including bioremediation, biofuels, and biomaterials.

In *Biotechnology in Medicine (BTC 1800H)*, the first of these two courses, students are introduced to leading edge developments in biotechnology with respect to their usage in medicine. Discoveries are traced from their inception in the laboratory through to the various phases of testing in patients ('bench-to-bedside'). The course focuses largely on therapeutics, paying particular attention to the recent application of biological compounds (or *biologics*), and to some of the hurdles faced by this newer generation of agents. However, diagnostics also enjoy some discussion. The goal of the course is to enable students to critically evaluate early-stage trial results, and to examine the consummate risks with respect to patient safety and business development. In addition, emphasis is placed on the value of appropriate competitor comparisons with respect to drug efficacy. More globally, *Biotechnology in Medicine* acquaints the students to some of the central challenges facing therapeutic development, and limits its scope to examining industry 'products' and issues that have commercial relevance within a five-year time-span. Some of the primary objectives of *BTC 1800H* include developing (i) a detailed understanding of clinical development issues, and the impact such considerations have on earlier pre-clinical development activities; (ii) the capability to critically evaluate press releases and product claims that cryptically communicate clinical and scientific findings outside the peer review process; and (iii) the ability to look at early-stage laboratory developments and evaluate or recommend possible paths for commercialization with a 5-year time horizon, with a strong emphasis on risk management.

The second of the two courses, *Biotechnology & Corporations (BTC 1810H)*, picks up where *Biotechnology in Medicine* leaves off by introducing students to some of the key aspects of the bio-commercialization process, again with especial emphasis on the healthcare sector. This course has its parallels with *Management of Technological Innovation (BTC 2030H*, see above), in that it seeks to harness together the students' growing understanding of scientific innovation (particularly with respect to new drug development) with their emerging business knowledge. However, while *BTC 2030H* approaches science and innovation from the direction of management, *Biotechnology & Corporations* approaches businesses from the scientist's perspective. In particular, the course focuses on the nature of corporate entities and the fundamental role they play in the development of new therapeutic drugs and diagnostic tools in the context of a highly regulated business environment. Topics covered include technology transfer, venture creation and financing, company analysis and evaluation, ethics and safety, and corporate structure and governance. The material is delivered by a wide range of recognized industry specialists, and the students benefit greatly from the opportunities to network with these guest speakers during the regular round-table luncheon sessions. As with other courses, the students are assigned to work in their teams to develop their first business proposal, which is submitted at the end of the course. Of particular note to this course is the high degree of preparation that is expected of students in order to accommodate the diverse subject matter.

Both of these hybrid courses seek to build complementary bridges that span the gap between the very distinct worlds of conventional academic science and modern technology-focused businesses. We have found that such courses are particularly effective for positioning the importance of scientific endeavors within the framework of management practices, illustrating the value of an interdisciplinary approach, and instilling in the students the kind of confidence with both disciplines that they will need to achieve truly differentiating success in the workplace.

STUDENT WORK TERMS

Aside from the carefully selected academic courses, one of the most attractive assets of the MBiotech Program from the perspective of incoming and prospective students is the well-established work-term placement scheme. During their time with the program, students are expected to participate in at least two four-month-long work-term segments (*BTC 1900Y, BTC 1910Y*) in local industry. The students are encouraged to choose from a wide range of internship roles on offer, and we are fortunate to be able to access a rich pool of positions in clinical affairs, marketing and sales in the biopharmaceuticals sector, most notably with GlaxoSmithKline, who have been a strong advocate of the program since its inception. Placements are also frequently offered by other large pharma companies in the region, and by smaller biotechnology firms. Nearly all such placements are located within the Greater Toronto Area (the conurbation of Toronto and its satellite cities), but some are located out-of-province, and we are actively encouraging and supporting students seeking further afield for their internships. Recently, we placed our first graduate student with an investment group based in Montréal, and facilitated the fulfillment of the individual's remaining academic credit requirements by offering courses by webcast, as well as recognizing enrolment in relevant course offerings at Concordia University.

Two of the work terms, *BTC 1900Y* and *BTC 1910Y*, may be fulfilled at a single organization, either in the same role or on two discrete four-month assignments, or they may be taken at two completely different firms. In addition to these two compulsory work terms, a third four-month term (*BTC 1920Y*) can also be taken as an elective. Historically, many of our students have chosen to follow this path, opting (at their employer's discretion) to extend their current internship either to eight months, or to a full twelve months, and thereby earning 2-3 full credits towards their degree, while continuing their placement with the same company. Others, however, prefer placements with multiple firms to broaden their exposure to potential employers. Regardless of the direction each graduate takes, our policy has been to be as supportive as possible, so long as the educational value is transparent for each of the three work terms. Thus, whenever a student chooses to remain with a single firm for an extended period, a process of continuous training is expected.

This system of placements offers students the opportunity for real-life learning in corporations, and provides economic incentives to companies, such as tax benefits, to recruit highly educated and motivated young people without exposure to long-term employer commitments. The interns are typically very enthusiastic and quickly adopt the culture of the host organization; and they have the chance to articulate their own individual value as team-players, and as potential employees down the road. The approximate, aggregate distribution of interns on these work terms, segregated by job function, is illustrated in Figure 2.

Growth of graduate programs has, in recent years, become the watchword for the continued success—and buoyant funding—of academic departments in Ontario. While we are now, akin to many of our colleagues running other Master's programs, under

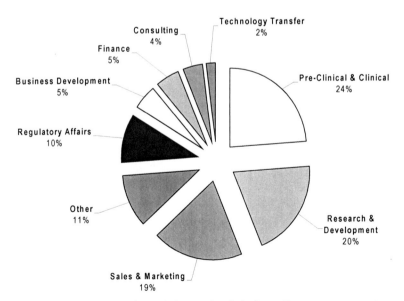

Figure 2: Distribution of work terms by job function

significant pressure to expand our graduate operations, we plan to do so cautiously, to protect and nurture both the MBiotech brand and the continued success of our work-term scheme. Consequently, it has become our first priority to ensure sufficient diversification of the work-term placements, such that the program does not become unduly behoven to any one sponsor. At the current time, a large proportion of our students are placed in clinical affairs and marketing positions within the large biopharmaceutical multi-national firms. During the next phase of MBiotech's development, our vision is of a gradual shift away from this bias, and we are actively initiating strategies to bring more of our students into the small-and-medium-sized enterprises for their work terms, with the aim of offering a broader spectrum of learning opportunities in the biotechnology sector.

BEYOND GRADUATION

The Master of Biotechnology Program began in 2001, and since that time has successfully graduated more than 100 students. Our first class of 14 students were each placed in a variety of companies for their work terms, in positions that ranged from marketing and clinical trials to research and development, and regulatory affairs.

Upon graduation, numerous students were offered positions in the companies where they had completed their internships. Several accepted these positions, and today they can be found still working for the same companies in increasingly senior positions. One student, in particular, began her career on a work term with the clinical affairs department of GlaxoSmithKline. She was initially appointed in a junior position and assigned a relatively unchallenging workload. After repeatedly proving herself, she was given more responsibility and eventually became a project manager for several clinical trials; all this while still a student in the MBiotech Program. This graduate from the Class of 2003 has continued to ascend through the ranks at GlaxoSmithKline, and already holds a mid-level management position.

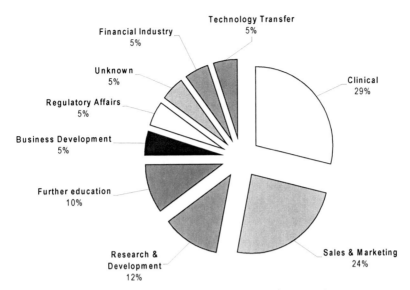

Figure 3: Distribution of post-graduate employment

Companies are not always in a position to offer our graduates full-time employment, and frequently students feel there may be a better fit with a different company, or a different department. Figure 3 summarizes the full-time positions graduates from the MBiotech Program currently hold after two-years in the field.

From this pie chart it is clear that the majority of our graduates are attracted to clinical trials (29 percent), as well as to sales and marketing (24 percent). A proportion of the graduates are drawn to research and development, but this historically accounts for fewer individuals, and includes areas such as process development and pre-clinical research. Approximately 10 percent of our graduates currently pursue further education in the form of medical, or law school. We have also seen some students continue on to PhD programs. At the present time, only one or two people graduating from the MBiotech have opted to enroll in an MBA Program. In summary, we can conclude that the MBiotech Program provides an excellent starting ground for students who have an interest in employment in the biopharmaceutical sector, and especially for those interested in project management roles.

FUTURE TRENDS

Our mission in the MBiotech Program has been to train today's innovative scientists to become tomorrow's business leaders. Although this mission has been specifically tailored for the student, we also now recognize a need, at the same time, to provide a learning forum for current professionals in the industry. The AstraZeneca public seminar series (*BTC 1600H, BTC 1610H*) is certainly an example of a milieu where both professionals and students may participate and learn. In this particular instance, this takes the shape of 26 weekday evening lectures coupled with live web-casts, for a full half-year of programming accessible to the general public, covering a wide range of topical subjects in biotechnology. However, our intention is build on this format by specifically reaching out to professionals with provisions for certificate-styled courses in biotechnology that our graduate students may, in addition, choose to take as electives. Two such courses are slated to run for the first time in early 2008. These bear the running titles *Medical & Scientific Challenges in Marketing Therapeutics* and *The Drug Discovery Process: From Bench to Bedside*. In both cases, the instructors are sessional lecturers drawn from industry, with the intention of fostering greater interaction between the university and the wider biotechnology community.

Biotechnology is an ever-changing field. As the twenty-first century unfolds, it is poised to rival—and perhaps supercede—the other major realms of human technological endeavor. The need for highly trained individuals to assist in the active commercialization of scientific discoveries and, consequently, for advanced educational practices is self-evident. As such, it is of paramount important that we continue to anticipate the needs of industry and adjust our curriculum to better serve the many facets of the biotechnology sector. It is our goal as teachers to remain at the forefront of educational innovation in this respect, and as we look to the future, we welcome the opportunity to develop and share best practices among our peers.

Columbia University M.A. Biotechnology Program
Carol S. Lin

Carol S. Lin, Ph.D., is Director of the M.A. and Postbaccalaureate Programs in Biotechnology at Columbia University. Carol can be contacted at CSL27@columbia. edu.

Columbia University's Free Standing M.A. program in Biotechnology[1] is an interdisciplinary program offered by the Graduate School of Arts and Sciences and the Department of Biological Sciences. It trains students in advanced science and related disciplines with a particular emphasis on approaches used in the biotechnology and pharmaceutical industries. The curriculum focuses on the basic principles of biotechnology and specific applications in various fields. The use of biotechnology to fight disease is emphasized. The program aims to prepare students for diverse careers in the industry such as research scientists in industry laboratories for drug discovery, in regulatory affairs for drug developments, and in investment or law firms that work with the biotechnology industry.

Thirty points (credits) of course work plus a master's thesis are required for the M.A. in biotechnology. The coursework includes three core courses, intensive laboratory experience and elective courses selected from the Department of Biological Sciences and a variety of other departments according to the student's specific interests. The thesis is a scholarly manuscript on subtopics in biosciences. The program can be completed by full-time students in one year including the summer term or at a reduced pace by part-time students.

---

1 *http://www.columbia.edu/cu/biology/pages/ma-biotech/pro/intro/index. html*

INCEPTION[2]

This program is designed for students who seek graduate level educations and a career in biotechnology without making the 5 – 7 year commitment to attain a Ph.D. The biotechnology and pharmaceutical industries are among the largest in the U.S. such that there is a great demand in these areas. Many career options require advanced training beyond a bachelor's degree but do not involve expertise acquired from Ph.D. training.

Prior to 2000, the Ph.D. program in the biological sciences regularly received requests from students for a master's degree, although its catalogue and web site explicitly stated that there was no free standing M.A. program. Further, an article in *Science*[3] reported that the industry were excited at the prospect of shorter master's degree programs which would provide employees with more relevant skills to their needs than Ph.D.s. Few universities in the U.S. offered such a program at that time. Therefore the Department of Biological Sciences decided to create one to fill this unmet need. A proposal was authored by faculty members from the department in 1999. The program received approval from all relevant departments and agencies and accepted its first class in September 2000.

COMPONENTS OF COLUMBIA UNIVERSITY'S M.A. PROGRAM IN BIOTECHNOLOGY

ADMINISTRATION

ADVISORY COMMITTEE

A supervisory committee consisting of 4-5 tenured Biological Sciences faculty members meets regularly with the program director. The committee establishes the structure and rules of the program, evaluates the program, and reports to the department

---

2   *Adapted from R. Prywes, Proposal for a Columbia M.A. in Biotechnology, unpublished*

3   *M. N. Jensen, (1999) Science Education: Reinventing the Science Master's Degree. Science, 284, 1610*

chair.

Program Director

The program director is a full time non-research faculty member. The director assumes the responsibility of day-to-day academic supervision and advising of students. This includes pre-admission consultation, academic and career monitoring, academic advisement, and career advisement for current students and alumni. The director teaches 3 core courses (strategy, seminar, and laboratory, see "curriculum" section) and coordinates research and thesis advisories. In addition, the director investigates new initiatives in curriculum development and establishes networking relationship within program, with outside departments in the academics and the industry.

Administrative Assistant

One part-time administrative assistant is dedicated to program affairs. In addition, departmental administrative staff offers assistance when needed.

CURRICULUM

Required

Core lecture courses
Thirty points (credits) of course work[4] plus a master's thesis[5] are required for the M.A. in Biotechnology. There are 3 required lecture courses: a technology course[6] where research strategy and techniques are emphasized; a strategy course[7] where drug discov-

---

4   http://www.columbia.edu/cu/biology/grad/biotech/electives.html
5   Guidelines: http://www.columbia.edu/cu/biology/pages/ma-biotech/cur/hndbk/thesis-guide.html
    Past theses: http://www.columbia.edu/cu/biology/grad/biotech/theses/theses_menu.htm
6   BIOL W4034 Biotechnology. http://www.columbia.edu/cu/biology/courses/w3034/
7   BIOL W4300 Drugs and Disease. https://courseworks.columbia.edu/cms/outview/courseenter.cfm?no=BIOLW4300_001_2007_3

ery and development strategies as well as vital oral and written presentation skills are covered; and a seminar course[8] where practitioners in the industry are invited to give weekly seminar and additional roundtable discussions. These three courses provide 9 points.

Laboratory or other internships

Laboratory work and internship[9] add 6 points to the required curriculum. Students either take an intensive summer laboratory course or conduct research within or outside of Columbia University.

Electives

The remaining 15 points (typically 5 courses) are chosen from a selection of over 70 pre-approved graduate level courses offered within Columbia University, from the Graduate School of Arts and Sciences, School of Engineering and Applied Sciences, Business School, School of Public Health, and School of International and Public Affairs. Students design their own program with the help of the program director and other faculty advisers. Of the 15 points, at least 9 must be from bioscience disciplines and departments such as biological sciences, biochemistry and biophysics, bioinformatics, genetics and development, pathology, pharmacology, or physiology. Some students focus their courses in specific area such as cancer research or bioinformatics. Others opt to expend their exposure to a diverse fields ranging from healthcare investment, pharmaceutical strategy, to bioethics.

Thesis

In addition to course work, a master's thesis is required. Most students choose to write a scholarly review of a subtopic of biotechnology. A smaller fraction of students have done substantial laboratory research beyond degree requirement. These students have the option of writing a research-based thesis. The third thesis option is to write a proposal suitable for submission for govern-

---

8   *BIOL G4305 Seminar in Biotechnology. http://www.columbia.edu/cu/ biology/grad/biotech/Seminar_Archive.htm*

9   *BIOL G4500-4503 Supervised Research. http://www.columbia.edu/cu/ biology/courses/g4500-g4503/SR_menu.htm*

ment grant funding, equivalent to NIH's RO1 or SBIR. To date few students have chosen this option, as such a project is often beyond the feasibility of those enrolling in a one-year program. However, one of the grant proposals has been submitted to NIH and is awaiting decision.

Thesis topics are chosen in coordination and with the approval of the program director. Students seek faculty advisors from academia and industry. Although students can choose advisors anywhere in the world with the consent from the advisor and approval from the program director, most students are advised by experts from academic institutions in greater New York area. Others are mentored by advisers in other U.S. states or in France, UK, and Israel.

Continuous curriculum assessment

The required curriculum is under constant evaluation and occasional revisions. For example, the seminar course was initially proposed to be similar to traditional science departmental seminars where academicians are invited to present a 50-minute seminar of their research findings. Within the program's first year, it was determined by the director, and agreed upon by the advisory committee, that these types of seminar do not offer unique learning opportunities to the biotechnology students. There are already numerous high quality seminars in the department, the university, and in New York City. The students need industry-related exposure which is not available in regular academic settings. Further, topics outside of laboratory research are of great interest. The seminar is now structured with industry practitioners of different subfields leading seminar and roundtable discussion. Invited speakers touch on topics ranging from non-profit science media, industry R&D, intellectual property, business development, to entrepreneurship. Most speakers divide their discussions into "industry trend" and "career development" sections. At the conclusion of the seminar, 3 – 4 students host a dinner with the speaker and continue their small group discussion and networking.

The other course that underwent significant transformation since program proposal is the laboratory requirement. Initially the curriculum was designed to require all students to take an intensive summer laboratory course. It was discovered within the first admission cycle that the class would be larger than an optimum lab course. Further, to the pleasant surprise of the program administrators, a large number of qualified applicants already have substantial undergraduate laboratory training, having done a year or more of advanced research projects. These students are encouraged to seek research experience in academic or industry research settings. A few students also come to the program with more than 3 years of research experience. These students seek career development outside of research arena. For them, another summer working in the laboratory does not add significant value to their training, and their laboratory requirement is therefore waived. In place of the laboratory requirement, they can choose to conduct an office-based project in law, business, or regulatory settings. The program director coordinates and oversees these laboratory and office internships.

OPTIONAL ACTIVITIES

In addition to required curriculum, students are strongly encouraged to participate in numerous optional activities. The program runs workshops in writing, presentation, and leadership skills, in addition to hosting networking events for career development. The program also actively disseminates information regarding relevant workshops and conferences organized by other institutions within the greater New York City area.

The M.A. Biotechnology students run a very active and successful student club, The Columbia Biotech Association[10]. The club is wholly run by the students, with advisory and sponsorship from both academia and industry. It seeks to provide information and opportunity to other graduate students, postdocs, and research associates from all institutions from the greater New York City area in their transition from academic settings to professional fields. The club has held many successful conferences and networking events. An additional benefit for the club is that it pro-

---

10 http://www.columbia.edu/cu/biology/BiotechClub/

vides opportunities for club officers to hone their organizational and leadership skills, as well as to establish personal connections with the invited speakers and sponsors.

Students are also encouraged to conduct independent biotechnology-related activities while enrolling in the program. Many students are actively involved in non-profit activities. Examples include co-directing a science communication consortium[11] and assisting in organizing a international medical assembly at the United Nations[12]. Additionally, the entrepreneurial spirits of the students has lead to the start up of three biotech firms with co-founders from the program and the university.

PEDAGOGY

KNOWLEDGE CONTENT

The strength of our program stems from the world class intellectual assets offered by Columbia University, and to a smaller but no less important extent from the world beyond Columbia. Students have the option to select from a very large number of high quality courses and activities lead by distinguished scholars and practitioners. They have the opportunities to seek advice from even more mentors who are knowledgeable and enthusiastic in offering highest level of intellectual exchange and guidance outside of classrooms. It is not rare that students decide to take more courses than required for the degree to better themselves for the expending knowledge required for a successful career, or simply for personal development.

TEACHERS

Our professors are not only knowledgeable scholars, they are also innovative teachers. With inspirations from the affiliated teachers college[13] and active faculty support from the Center for

---

11 http://scicommconsortium.com/home

12 http://www.medicalassemblyun.org/

13 http://www.tc.columbia.edu

New Media Teaching and Learning[14], the professors use diverse strategies and technologies to engage the students and foster the best advancement in learning and thinking.

LEARNERS

Students are our most important assets. Our strength and reputation enables us to draw a large number of highly qualified applicants from around world including Canada, Taiwan, India, China, UK, France, Japan, Korea, and many other countries. We can afford to carefully select the best match based on applicant preparation and aspiration. Once we admit a student, it is our mission to educate and mentor to the highest possible level.

Our students have diverse backgrounds. Approximately one half of the students come to us directly after college graduation, majoring biology-related fields. The other half join us after they have established careers, ranging from a few years to reaching retirement. While all our students have biology background, some have additional graduate degrees as MA, MBA, JD, or MD, prior to joining us. They are all interested in developing a career that utilizes their knowledge in bioscience. The great benefit of such a diverse student body is that they not only learn from their mentors, they learn from their peers.

Our students are incredibly intelligent, hardworking, and savvy. Since all our classes and activities emphasize interactions and team works, they also develop strong camaraderie. The friendship established in school often extends beyond graduation, resulting in long lasting personal and professional relationship.

MILIEU

Columbia University and New York City offer an unsurpassed learning and networking environment. As Columbia's mission statement indicates, it is one of the world's most important centers of research and a distinctive and distinguished learning environment. The financial leadership of New York City, the location

---

14 http://ccnmtl.columbia.edu

of many pharmaceutical company headquarters in the greater New York area, the City's status as second-largest recipient of NIH funding[15], and the area's status as largest biotechnology patent holder in the U.S.[16], all contribute to an unmatched environment for the students to interact and develop into the best biotech practitioners possible.

ENABLING ENVIRONMENT

Supportive higher administration, excellent infrastructure, state-of-the-arts computer facilities and laboratories, extensive library collection and online access are all vital to the program's success. Most importantly our mentors and advisors are the most crucial enabling factors. These include faculty and staff within the Department of Biological Sciences, other departments within Columbia University, other academic institutions in New York City and beyond, and practitioners in the industry. Whenever the program reaches out and seeks help or advice, the response has always been overwhelmingly positive. Our advisors and mentors spend time and effort with us; sometimes in a 10-minute phone conversation, sometimes in a semester-long course. They take our students under their wings and show them the tools of the trade. Our program could not have succeeded without the enthusiastic and generous support from people outside of the program.

OUTCOME ASSESSMENT

To date, we have 149 graduates from the program's 6 year history. Columbia's biotechnology alumni have careers in medicine, laboratory research, regulatory affairs, clinical research, intellectual property, business, and education. Examples of our alumni's careers are: staff biochemist, Merck Research Labs; clinical research scientist, Novartis; senior director, WebMD; attorney, Johnson and

15 $1.3 Billion received in 2003. Data obtained from New York City Economic Development Corp.

16 6,800 patents granted from 1990 – 1999, compared to San Francisco metro area's 3,991 and Boston metro area's 3,007. Data obtained from New York City Economic Development Corp.

Johnson; psychiatry resident, University of Pennsylvania; director of investment, Singapore Economic Development Board; associate vice president, business development, Paramount BioSciences; founder, BioNano Systems; postdoctoral fellow, Columbia University; and, Ph.D. student, UC Berkeley.

We continuously evaluate the program to appraise its alignment with our vision and mission. We take surveys from our graduates. We consult with industry experts. We ask for their frank assessments and recommendations. We make changes when necessary based on their feedbacks. We thrive to achieve the goal set forth in the university's mission: "to advance knowledge and learning at the highest level and to convey the products of its efforts to the world."

Biotechnology and Biomanufacturing at the Community College of Baltimore County: Integrating Research and Production

Thomas J. Burkett

Tom Burkett, Ph.D., is Associate Professor and Biotechnology Program Director at The Community College of Baltimore County. Tom can be contacted at tburkett@ccbcmd.edu.

Baltimore sits near the southern terminus of the most concentrated bioscience region in the country. Extending along the eastern seaboard from New England in the north to Northern Virginia in the south, the Northeast-Mid Atlantic region is host to a diverse bioscience industry base consisting of federal research laboratories, academic research centers, start-up companies, mid stage companies, large scale manufacturing operations and traditional pharmaceutical companies, along with all of the ancillary support products and services. Within the state boundaries, Maryland recapitulates this bioscience diversity, providing a multitude of career opportunities.

The challenge for educational institutions seeking to serve such a diverse industry is to design and deliver programs that meet the core needs of the industry, as well as the needs of students. Biotechnology training programs have existed since the early 1980s however—along with the industry—they have changed. Initially, university courses in molecular genetics gave way to post baccalaureate technical training programs, followed by the creation of the first technical and community college based programs, and now the ubiquitous professional masters programs that serve to meld the science and business aspects of this industry. Accompanying these academic courses have been a profusion of short courses, often provided through industry organizations, and specialty consulting / training companies

Community colleges have a long history of providing support to local industries through both general academic training as well as industry specific workforce training. Typically these two separate but complimentary functions have been offered through a mixture of credit and noncredit courses. At the Community College of Baltimore County (CCBC) we have used a mix of both credit and noncredit courses to meet the needs of our local biotechnology industry, as well as prepare students for transfer to four year university programs.

The Biotechnology and Biomanufacturing academic program at CCBC offered its first courses in the fall of 2000. Established with the assistance of a National Science Foundation Advanced Technology Education grant, the program has evolved from its early focus on academic research skills to encompass both research and production aspects of the industry. In this chapter I describe the program, the educational approach taken, as well as some of the program adjustments that have been made along the way.

THE ACADEMIC PROGRAM

When asked to describe the program at CCBC, I often state that it is project oriented, product focused, and overwhelmingly weighted toward laboratory skill development. The objective of the program is to produce research and manufacturing associates who have an understanding of the basic techniques of recombinant DNA manipulation, protein expression, and protein purification. Our courses use the development and production of biopharmaceutical proteins as a common theme to explore laboratory techniques and conceptual knowledge needed in this field, including the regulatory and quality aspects of producing a therapeutic protein.

STUDENT PROFILE

Community Colleges, through their open enrollment admissions policies and focus on workforce development, attract a dizzying diversity of students. The diversity of our student body is reflected in the students enrolling in the Biotechnology and Biomanufacturing Program at CCBC. Although there is year-to-year variability, what we have found after several years is that about a third of our students are typical high school graduates, about a third already possess college degrees—some with advanced degrees—and the remainder are older adults already working in or attempting to enter the industry but with limited experience in the laboratory. Graduates from our program have found employment in local companies as laboratory technicians, manufacturing associates, media preparation technicians, and cell culture technicians, among other titles. Both student demographics and the needs of our local industry were taken into account while designing the curriculum as well as the logistics of delivering the curriculum. Although the focus in our curriculum is on laboratory skills, sufficient academic content needs to be included so that students will be able to transfer credits obtained at CCBC to a four year university.

THE INTRODUCTORY COURSE

The Biotechnology & Biomanufacturing Program at CCBC currently consists of three core biotechnology and biomanufacturing courses, an external internship program, and supporting courses in microbiology, chemistry, and genetics. The program was initially designed around two core biotechnology courses focusing on nucleic acid and protein technology. However, it became apparent in the first year that an introductory course focusing on basic lab skills, vocabulary, and concepts would be needed before more advanced techniques could be introduced in the later courses. This course, our introduction to biotechnology, was incorporated into the program during the second year of the program.

Although students entering the program have taken freshman-level majors courses in biology and chemistry with laboratory, the laboratory skills taught in these courses did not seem to be an adequate preparation for the biotechnology courses. Laboratory skills such as solution preparation, pipetting, and aseptic techniques were inadequate or nonexistent. We therefore designed our introductory course to focus on these core skills, and to provide an initial exposure to molecular biology and protein purification techniques in the context of biopharmaceutical development and production. Topics such as quality and regulatory aspects of biopharmaceutical development are also introduced during this introductory course and expanded on in later program courses. However, unlike our later courses which focus on laboratory projects that span the entire semester, the laboratory offerings in our introductory course are independent from one another and more closely resemble the typical laboratory format with pre-lab work, laboratory, and post laboratory reports. The format for our later courses is quite a bit different, with more responsibility placed on the students. The more traditional format of the introductory course also serves to ease the transition from a typical structured laboratory environment to one progressively less structured and more dependent on student initiative and group work.

PHILOSOPHY OF CURRICULUM DESIGN

I had several guiding principles in planning the lab activities for what had initially been our first biotechnology laboratory course.

First, the course should have a project focus. The experiments in the course should be incorporated into a laboratory project so that students see and experience how independent techniques are linked and utilized to complete a project. For example, instead of doing an isolated technique such as PCR with some previously chosen primers and template that is virtually guaranteed to work, the students should design their own primers, isolate their own template, and figure out the PCR conditions necessary for the chosen application. This means doing a lot more PCR reactions, and running many gels. But at the end of the semester you have confidence that the students understand PCR and can tackle a PCR project without much assistance.

Second, the course project should address core skills of cloning and protein expression and in some way be oriented toward producing a "product" such as might be expected in a biotechnology company.

Third, no kits! The use of kits is ubiquitous in today's bioscience laboratory and I do believe that students should be familiar with the various types of kits available for doing plasmid preps, RNA isolation, PCR and so on. However, I also wanted to stay away from the exclusive use of kits, believing that being able to assemble the reagents necessary for a technique would provide more experience in those critical fields of solution preparation. Also, in constructing solutions from scratch one has the opportunity to address the different solution ingredients and their function. Being familiar with the individual ingredients and what they do will help students later on in troubleshooting experiments that have gone wrong. In addition, I believe that the use of educational kits, while helpful and necessary in some settings, stifles innovation and exploration in the teaching laboratory. If we as educators are to be fostering these attitudes, then I feel we should be modeling them for students. Involving students in a project that the instructor is intimately involved with will help the students see the larger questions as well as some of the underappreciated details of a project, whereas simply following the directions in an educational kit limits creativity and can obscure the nuances of a given experiment that can provide valuable learning experiences for students. Besides, troubleshooting experiments is a very effective way to teach critical teaching skills. Not using educational kits does create more work, however I think it allows more creativity in the classroom.

Finally, students need to work together and need practice in talking about, and writing about science. Often this is accomplished through oral presentations, or poster sessions. Lately I have been incorporating a class "wiki" the goal being the creation of a collaboratively written "paper" that describes the student projects and includes the data from those projects.

BIOTECHNIQUES I

In contrast to the introductory course, the *Biotechniques I* course taught at CCBC focuses more on the types of skills that students might need while working in an academic laboratory or small biotechnology company: (primarily microbial) cell culture, nucleic acid isolation and analysis, cloning, and PCR. Instead of laying out the laboratory exercises in the cookbook format as is usually done, and is also done in our introductory course, *Biotechniques I* teaches students to develop their own methods, protocols, and SOP's based on the scientific literature. Extensive use is made of web-based resources, including the National Center for Biotechnology Information[1] as well as company (NEB, Promega, Invitrogen, etc.) web sites and other molecular biology tools available on the web (primer design, etc.).

The laboratory section is based on a scenario of product development. Students, now staff scientists at "Biotech R US," are tasked with developing a product that has been selected by the management. Our most recent product has been the DNA polymerase I protein, otherwise known as the *Taq* polymerase, from *Thermus aquaticus*. The students start by researching sources of the bacteria, cultivation conditions, and any other available information on the gene and protein (such as cloning, DNA / protein sequence, and expression techniques). As we progress from media formulation to cultivation and DNA isolation, we start discussing cloning methods including the use of PCR. Eventually the students (with faculty guidance) develop a strategy to clone the *T. aquaticus poll* gene using PCR. The students employ web-based tools, such as PCR primer design programs, to identify suitable primers after lecture sessions covering the principles of primer design, stability, length, composition, hairpin, dimer, etc. The students then develop a few primer choices (which are ordered according to the student specifications), and many PCR reactions later usually get a product of the right size. Along the way they puzzle over controls, PCR programs, the use of DMSO, primer and template concentrations, and in general go through in a condensed and guided form what

1   *www.ncbi.nlm.nih.gov*

R&D scientists do on a daily basis.

During the course of the semester a really interesting transition usually occurs after about the third failed PCR reaction. Some get depressed because the experiments are not working, but what has happened more often than not is that the students stop working separately and start working collaboratively to get the project done (not all classes do succeed in their "project"). The students take ownership of the project and this, more than anything else, contributes to the success of the class. I have tried to figure out what causes this ownership to emerge, and from my limited experience there are two things that seem to foster it. One factor promoting project ownership is the course wiki. A wiki is a method for collaborative writing and sharing information, the best known example being Wikipedia—a collaborative encyclopedia where anyone can edit and modify entries. The class wiki is set up so that members of the class can make entries, post comments, or modify entries. The goal is the collaborative documentation of all the laboratory experiments, methods and results written in the form of a scientific paper. Initially, the course instructor walks the class through collaboratively writing the various sections, including any relevant references. However, students eventually take over the wiki. The wiki then serves as the background information for a final oral presentation which is open to the public. The other facet which seems to foster project ownership is focusing on the students solving the problem: provide them with the tools and a little guidance and they can come up with the conditions to get their "project" to work.

BIOTECHNIQUES II

Our final course in the biotechnology curriculum, excluding a required internship, is a little more structured then the previous course, *Biotechniques I*. Ideally students take *Biotechniques I* before *Biotechniques II*, however on occasion students take *Biotechniques II* prior to taking *Biotechniques I*—it is not ideal, but in the real world of maintaining enrollment in these specialized courses sometimes exceptions have been made.

In *Biotechniques II* the focus is on the production aspects of biotechnology: proteins, protein expression from recombinant sources, and purification. Industrial aspects such as quality systems and regulatory requirements were introduced in the introductory course and here they are expanded upon in the form of raw material specifications, quality control assays, and production batch records. Since we are dealing with proteins that could be therapeutic entities, cGMP considerations are always present.

As stated earlier, *Biotechniques II* is a bit more structured than *Biotechniques I*, since the goal is gaining experience with techniques that would be used in both research and production environments. Several microbial expression systems, including *E. coli*-synthesized GFP, *S. cerevisiae*-synthesized GFP, *P. pastoris*-synthesized HSA and ß-Gal, are utilized to illustrate aspects of protein expression including operation of bench scale fermentation units, concentration and clarification techniques, and liquid chromatographic methods for purification of recombinant proteins. The class will typically go through three guided expression and purification exercises using the above systems. However, while doing so they are also working on the expression and purification of their "project" protein carried over from the previous semester. As in the previous semester, the class develops the project including all of the production documentation that would be utilized including QC specifications.

CONTINUING EDUCATION AND BIOTECHNOLOGY

The Biotechnology and Biomanufacturing Program curriculum at CCBC was not developed in isolation, but with the input of several industry representatives. One of the insights gathered from numerous meetings with industry was the need for short courses that would teach critical skills and topics but not be constrained by the academic calendar and credit class structure. Working hand-in-hand with the continuing education division of the college, state, and county economic development officers, and private training companies, a consortium of training providers was formed to address the need for short courses. Surveys of industry needs were conducted by economic development officers and from the results of those surveys several courses were developed and marketed to the local industry. These courses were developed as focused one to two day offerings catering to specific needs of our local industry. Courses offered to date have included Aseptic Processing, introductory and advanced courses on cGMP, basic laboratory documentation, and change control. Additional course offerings based on industry input are currently being developed for offering by CCBC and the other members of our consortium.

ESTABLISHING COMMUNITY COLLEGE BIOTECHNOLOGY PROGRAMS

Community colleges have an important, if not critical, role in biotechnology education. However, developing and sustaining these educational programs can be a Promethean task. In most cases the majority of the burden in establishing these programs is borne by one faculty member often with encouragement, but little tangible support, from the administration. Fortunately, a number of organizations including *www.bio-link.org* and *www.biomanufacturing.org* have been funded by the National Science Foundation in order to develop and share instructional resources as well as advice on establishing and running community college based biotechnology and biomanufacturing programs. Both organizations offer workshops and conferences that can get a person developing a biotechnology program at his or her local community college up to speed in a hurry.

The establishment of a community college-based biotechnology or biomanufacturing program is a difficult, but rewarding, task that can be made easier with the help of the organizations named above.

Other considerations that should be taken into account as these programs are developed include:

- The nature of the local industry and their hiring needs. If the local industry is dominated by a large manufacturing program, then that will influence the design of your program and curriculum; localities focused on agricultural biotechnology will have a different program structure and curriculum then areas dominated by pharmaceutical manufacturing.

- The recruitment of an industry-based advisory board that is willing to provide jobs for students and training for faculty members. Continued skill training of faculty is important, especially concerning the industrial aspects of biotechnology.

- Will there be a dedicated space for the biotechnology laboratory? A dedicated space is important if the program is to accumulate the equipment and reagents necessary for conducting experiments.

- Will there be laboratory support such as for laboratory technicians in the courses.

- Funding! The initial cost of these programs can be considerable, with substantial up front equipment and reagent requirements. I would encourage the use of "home-made" reagents instead of educational kits. There are educational reasons for favoring home-made reagents, but there are also logistical reasons as they can reduce operational costs and provide flexibility in experiment design.

- Incorporate "soft-skills" into your courses. Employers often list team work, communication, and critical thinking as the most desirable skill sets.

- If a full scale academic program is too ambitious and not warranted by the local industry, consider the development of short courses offered through the continuing education division of your school.

The Role of the Regional Biotechnology Center: The Development of an Emerging Convergent Technology as Curriculum

James Harber

James Harber, Ph.D., is Director of the Central Coast Biotechnology Center at Ventura Community College, James can be contacted at *jharber@vcccd.net*.

A quick study of the biotechnology grants infrastructure for the United States suggests that there are significant challenges in new technology introduction. Deployment involves new education of faculty, integration into existing text based laboratories and maintenance of the new technology program. Biotechnology upgrades are particularly challenging because the technology to be introduced may simultaneously involve new convergent knowledge of several disciplines, including biotechnology, nanotechnology, computer science and cognitive science.[1] This chapter reviews the development of a biotechnology program from its origins as single technology classroom modules taught by industry instructors to a statewide grant funded organization with many centers.

The responsibilities of biotechnology center directors include encouraging individual colleges to maintain and expand their curriculum for training students for employment in companies and transfer to the university. Additionally, a valuable service to existing biotechnology employees and executives is provided by a center offering skills upgrades and specialized training upon request. Center directors also devote a great deal of time to networking with economic development professionals and the venture capital

---

1   *Managing Nano-Bio-Info-Cogno Innovations; Converging Technologies in Society by W. Bainbridge and M. Rocco, 2006*

community where appropriate, as these resources provide key indicators of the ongoing emergence of new technology companies, and spin-offs of existing biotechnology companies. In California, there are currently six centers for biotechnology, two hubs, and a statewide director. More centers are planned. The entire effort was the brainchild of two key faculty-administrators at Ventura College (Dr. Robert Renger and William Thieman, Dean and Chair respectively at Ventura College) and developed to become part of the California Community Colleges Workforce and Economic Development Program.[2] In addition, a statewide grant writing team organized by Robert Renger pulled together national resources to launch the Bio-Link Center in San Francisco.[3]

Those of us who have been concerned with the practicalities of biotechnology program educational development for entry-level employees in start-up companies know the challenges of acquiring and introducing new curricula. Often an opportunity starts with a new networking event and a new conversation. Quickly communicating the essence of biotechnology has been vital. An interdisciplinary definition of biotechnology that satisfied both educators and venture capitalists to the following: *You make loads of money in biotechnology by brewing genetically engineered wine in swimming pool sized stainless steel vats while wearing a space suit.* This single sentence translates the mundane fundamentals in biotech commonly known as adequate cash flow, high technology and skilled technical labor into an easy to remember phrase, "making wine in space suits." Inventing the language and allegory for bringing together different disciplines for biotechnology educational program development involves a rather basic algorithm of activities that is repeated each year with the grant cycle. The methods described in this chapter summarize the essence of the lessons learned from seven consecutive grant cycles by a center director and suggest a common algorithm that can be reapplied as needed to biotechnology public-private partnership efforts elsewhere.

---

2   *www.cccewd.net*

3   *www.bio-link.org, also discussed in this volume*

THE CENTRAL COAST BIOTECHNOLOGY CENTER

The story of economic development in biotechnology at the community colleges of California anecdotally starts in an unlikely sphere—with the animation industry. As computerization and gaming developed in the 1980s, digital alternatives to hand drawn cartoons were clearly in demand. The problem was that computer technical education had not kept pace, despite the fact that the computer industry was being invented in California at the time. The result was that the average entry-level digital animator at the time was educated out of state. Ironically, there was also the paradox of significant unemployment in California. The animation industry petitioned the California legislature for support for technical training. This led to the establishment of a statewide economic development network (ED>Net) at the community colleges in the 1980s which later became the Economic and Workforce Development Program (EWD) in the new millennium.[4] Biotechnology was added to this statewide mandate in the mid-1990s. Robert Renger, Bill Thieman, Kim Perry and Mary Pat Huxley have been successive statewide directors for the EWD initiative now know as Applied Biotechnology Centers. Collectively the six regional centers and statewide coordinator are collectively funded with roughly $1.5 million from the EWD. It is expected that center directors seek external funding beyond this seed funding for technology curriculum development.

In the case of the Central Coast Biotechnology Center (CCBC), resources were acquired and generated in the past seven years through partnership with biotechnology companies Amgen, Baxter, Ceres, Dako, Integrity Biosolutions, Invitrogen, Alliance Protein, Alzheimer's Institute at CSUCI, Fziomed, Hardy Diagnostics, Coastal Marine Biolabs, Technical Associates, and Promega, During this time frame, collaborative grants were arranged with Amgen (3-D projection), California State Universities at Channel Islands and San Luis Obispo (grant development and review), College of the Canyons (126 Biotech Corridor), and

---

4   *www.cccewd.net*

UCSB (California Institute of Regenerative Medicine – stem cells). Despite these intensive efforts, the acquisition of an emerging transformational convergent biotechnology which would fundamentally alter basic college instruction and simultaneously provide industry training benefits was elusive and would wait until further collaborations developed as described below.

As an aside, it is important to note that other allied technical training initiatives, (multimedia, advanced transportation, competitive technologies, international trade, workplace learning, and health) are among the palate of EWD initiatives. The EWD mission in brief is to advance California's economic competitiveness through a well-trained workforce. However, the method(s) used to achieve this aim by the CCBC are worth noting as they can provide significant direction to those with similar challenges. Implicit within the algorithm for biotechnology public-private partnership is the goal of describing the basic methodology for generating resource development (equipment, material and funding) to augment the meager support for high technology education.

THE CCBC CORE ALGORITHM

The CCBC has operated for over a decade with a core algorithm that starts each year by renewing partnerships with businesses and key industry advisors early in the fall. Because the center operates during the summer months, it is prepared for the regular networking events that take place in business circles starting in September. Educators typically underestimate the importance of informal regional business meetings and likely don't have the time, or have scheduling conflicts with attendance. In California there are many professional opportunities for networking, including the MIT enterprise forum, Biobrew, IEEE, and regional biotechnology organizations (SoCalBio, BayBio, Biocom). Other avenues for networking include small business innovative research meetings (SBIR) and trade shows which offer combinations programs for investors and scientists (Biotechnology Industry Organization, LARTA). In contrast to technical meetings common at universities, these industry organizations provide the opportunity to meet

The CCBC Core Algorithm

1. Renew a broad set of partnerships annually.
2. Informally / formally survey businesses, faculty, and individuals.
3. Define regional needs with a core of advisors.
4. Strategically address the need and incrementally improve the regional curriculum.

a wide diversity of people from different fields, including representatives from the U.S. Patent Office.

Beyond the networking aspect of the algorithm for generating public-private partnerships is the practical need to generate a core of advisors. Face to face meetings to advise a biotechnology education program are not as important as the individual relationships that a center director develops. Also, in keeping with the high-risk business model, many collaborative ideas with advisors fail, but this should not impede reformulation and refinement of the basic collaboration with the representative from industry. The main goal of collaboration with an advisor from industry is to find out what type of employee that person would hire. Then, implement at least one technology training module for that hiring need in a classroom, independent study, or workshop. This involves considerable finesse when dealing with faculty across a region. Again, since this is done on a yearly cycle, the retirement of unneeded or less desired aspects of the training curriculum is critical. This opens the grant deliverables and makes readdressing of industry needs each year possible, with new deployment modules. Many studies have been written on common skill sets needed for industry. The best way to find out what training is needed locally or regionally is to ask your advisors directly. Using this method, exceptional lab modules were developed for atomic force microscope, formulations, protein stability, cell cultivation, bioreactor, and fluorescent protein expression in prokaryotes and eukaryotes by transfection and virus infection. Offsite modules in HPLC, structural biology, and other labs involving extensive instrumentation were performed at industry sites using their equipment.

The next activity in the yearly algorithm is formulation of a survey. Major surveys require a great deal of time and money and should only be undertaken when a major shift in the program or local industry need is expected. In the case of the CCBC, a major survey was undertaken in 1998, prior to Amgen's massive expansion, Baxter's opening of a biosciences manufacturing suite, and Biosource's (now Invitrogen) extensive new hiring. The chapter on the Moorpark Program, elsewhere in this volume, illustrates a training program developed from specific industry polling. Minor surveys can be as simple as a quick phone call to human resource departments and lab support companies, assessing which jobs are in demand.

Most hiring managers realize that within a class of students, they will probably only want to view several of the best resumes. For this reason, the center director encourages educational modules such as resume writing wherein a company in need of employees can come and teach a section on the fundamentals of preparing documents specific to their company.

Beyond minor and major surveys is the need to coordinate local needs with statewide or national employment need trends. For this reason, the CCBC became involved in the California Governor's life sciences initiative that surveyed all major biotechnology companies, educators, venture capitalists and economic development/labor specialists simultaneously. Four documents were produced which collectively define a roadmap for biotechnology economic development in California. These documents predict that business formation and research growth will occur primarily along the north-south coastal belt, while manufacturing and expansion will take place along the east-west corridors that connect the coasts with the central valley and deserts. These crucial documents can be found at *www.monitor.com* and are closely linked to articles from Michael Porter on technology cluster model theory. In brief, the cluster model suggests that new emerging startup companies will have the greatest entry level training needs (like animation in the 1980s). These startup companies grow more frequently near larger existing companies—during times of distress in the larger company, employees who want to stay in the region typically start their own ventures near the parent company. These small companies have special needs that can be met by small community college training programs.

One of the great success stories of CCBC is that Ventura College placed some of the first employees at Amgen, when it was a tiny company in what is now known as Building 1. In the 1980s, Amgen's fate was not assured and each hire was critical. At that time a former Ventura College student named Timothy Osslund became one of the first employees. There was no university in Ventura County at the time so companies like Amgen relied on relationships with community college programs for entry-level employees. This early partnership illustrates a judicious use of resources by the college. In the 1980s, Ventura College began offering biotechnology modules as part of regular college courses. This led to full program development later in the 1990s.

New technologies have specific training needs that entail bringing faculty into the company for apprenticeship and training. This was encouraged during the mid-1990s and allowed faculty to return to the classroom with an appreciation for the goal of introducing specific curriculum into existing courses. Many of the skill sets are now taught as part of a research programs at the university. As community college programs with biotechnology faculty develop certificate programs and articulations with universities, the need for annually renewing partnerships and defining new needs is vital. For this reason, it is suggested that like Ventura College, schools elsewhere should offer courses which provide the public and non-majors the opportunity to hear a new lineup of biotechnology related speakers. The second key aspect of the Ventura College program is a lab skills course that affords those with a basic biological sciences background as well as those with a university degree (and no lab skills) the opportunity to learn lab technique directly from industry employees. Many of the instructors are former students. The founder of the program, William Thieman, is currently publishing the second edition of a biotechnology text used in the program. Terry Pardee continues to lead coordination of the industry sponsored laboratory instruction course he authored with considerable assistance from Marta deJesus and occasional input from the CCBC director.

Courses where students, faculty, and industry employees interact are the perfect environment to try out new curriculum ideas. The center director can develop new courses to test deploying emerging technologies, either by inviting industry directly to the classroom, or vice versa. In the case of the emerging curriculum described below, a representative from Cepheid flew into town with a suitcase one evening and spent an hour demonstrating the portability, convenience, and speed of a new portable real-time PCR (RT-PCR) product—all attributes that were highly sought in developing transformational curriculum modules. Additionally, the technology was built in California, satisfying the EWD mission to strengthen global competitiveness of local products.

The regional goal of the CCBC is to foster development of biotechnology at all the community colleges in the proximal coastal region. Because the current CCBC director had also participated in the development of the Moorpark Program in its first semester in the lab at CSUCI (with Marie Panec and Dr. Maureen Harrigan), there was awareness that it had considerable internal inertia and was highly capable of guiding its own development. The contribution of the CCBC to Moorpark has been an on-call relationship to provide resources when needed and to participate as an occasional lab instructor. The CCBC director contributed several chapters to the Moorpark-NSF Industrial Biotechnology text, particularly to encourage testing for *Mycoplasma* by PCR. Baxter quickly adopted this training locally in its QA/QC processes for producing factor VIII for hemophilia.

From 2002-2007, the CCBC algorithm of searching for business needs was extended to community colleges throughout the Central Coast region from San Luis Obispo to Ventura and eastward to Lancaster. This assigned region followed the state of

California economic development map. Partnerships were developed or renewed with community college science programs and biotechnology-related businesses at various stages of the CCBC project. The CCBC regularly interacted with regional universities, including Cal Poly SLO (for review of faculty grants with Susan Opava), UCSB Department of Neurosciences (for stem cells infrastructure development with Dennis Klegg), and the newly developing CSUCI in Ventura County (for the business and technology partnership committee and for the Master's Program in Biotechnology with Wayne Davies and Dr. Ching-Hua Wang among others). Each of these partnerships with colleges and universities involved layers of contacts from industry and education. A significant effort was developed in partnership the Gold Coast Innovation Center (for alternative energy) which resulted in a pilot biodiesel plant at the Hueneme Naval Base in Ventura and a $4 million biogas facility built by Onsite Power at UC Davis. Collaborators for alternative bioenergy included the National Park Service, County of Ventura, VCEDA, NFESC (U.S. military), Biodiesel Industries, and CCBC.

Perhaps the most fortuitous partnership the CCBC leadership inherited from predecessors at Ventura College was with a nonprofit organization, VCBio, that attracted the interest of a successful businessman and investor, Gary Clark. VCBio functioned under Clark to network the biotechnology interests of the Ventura, Santa Barbara, and San Luis Obispo from 2001-2004 and resulted in many industry-training partnerships. Unlike traditional academic program development, this training was directed at industry representatives. The discovery of this executive need led to the closure of the full cycle of the algorithm described in this chapter. Gary Clark now leads a local chapter of Tech Coast Angels, the largest investor in new companies in Southern California.

A difficult aspect of the CCBC algorithm as described (which correlates to a grant workplan) is understanding where to initiate the effort to upgrade the curriculum with new technology. Over the years, various efforts have included a vast number and spectrum of individuals in combination with technologies. The core question that continued to be asked was: *How is the center going to acquire a truly transformational emerging technology and succeed in its adoption throughout the college curriculum?* Through careful one-on-one surveys information it was discovered that many Amgen entry-level research technicians had been trained in microbiology lab skills at a community college prior or during employment. The CCBC then focused on the ubiquitous microbiology laboratory class as the common denominator to introduce a universally applicable curriculum.

FOCUS ON TRANSFORMATIONAL TECHNOLOGY

The CCBC realized that if the technologies were the emerging part of the economic equation that related to the EWD mission, it could simply pick a transformational technology that met specific criteria (speed, portablility, convenience, computerization, sensitivity, genetic-basis) and train everyone with it. The answer came from an unlikely source in a partnership with a geographic information system (GIS) grant authored by William Budke. The CCBC was asked to consult on detection of bacteria for a GIS project for High School Students. Initially, gel boxes, PCR machines, and other aspects of traditional methodology were considered. Then the labor was analyzed. The existing technology would involve constant monitoring, mentoring, and an ongoing commitment to interpreting bands on gels results. This existing technology would require intense time commitments. As described previously, the CCBC arranged demonstration of a specific implementation of portable RT-PCR product manufactured by Cepheid. The GIS project required detecting and mapping bacterial contamination in streams, which could lead to beach closure. This is a huge seasonal problem in California. The RT-PCR technology passed the initial review for inclusion in the grant, which was ultimately approved. The project seemed to be in trouble because the steep price of the technology meant that the purchase required college district board approval. With a one-year grant cycle, months of waiting can spell disaster for workplan timelines. Program officers monitoring the grant become impatient, and in this case there was also pressure to show grant progress before the state governor's election in November. In this case however, the real tragedy was that the spinach – *E. coli* 0157:H7 outbreak in Salinas killed and injured over 50 people, many of them children some of whom died. That hit the owner of the spinach packaging plant, Westlake, CA-based Dole foods, very hard. The Cepheid technology was used to track down the source of the outbreak in partnership via a New Mexico testing lab. This point was broadcast all over the news. A key question was, *Why was a California technology was being deployed out of state to solve a problem at home?* It harkened of the problems experienced decades earlier in the animation industry.

In September, the real-time PCR machine arrived, months behind schedule, but in time to contribute to a very relevant story. The technology was very sturdy—it had been developed on a DARPA SBIR grant—and was deployed in over 1000 post office sorting areas for detection of anthrax following the often forgotten letter incidents after 9/11. In fall 2006 the real-time PCR unit, being portable, was put in a suitcase and traveled all over the state for workshops. The reagents did not need to be refrigerated and, using a simple of the protocol, the product was able to detect the genetic signature of specific bacteria in as little as 12 minutes—in contrast with conventional bacterial identification methods, which can take experts 24-48 hours or more. This real-time PCR technology was portable and could also be used by a nurse in the hospital to swab a new mother for a bacteria (GBS) that could potentially cause meningitis in her newborn. Alternatively, screening all incoming hospital patients for the dangerous MRSA bacteria is now feasible. The inventors also implemented FDA level stringency to tests performed with the units. Furthermore, transforming basic microbiology laboratory curriculum from the existing Koch-Pasteur methodologies to a contemporary 21st century genetic signature methodology seemed to be feasible. The curriculum adoption of this key Nobel Prize-winning technology after so many years of waiting finally seemed feasible. It was easy to use; simpler than a cell phone.

The CCBC took every opportunity to demonstrate the Cepheid RT-PCR technology. Often the audience would consist of people without lab experience who would be fascinated that they could accomplish an advanced molecular biology experiment within an hour. After years of pushing projects along, one module and one instructor at a time, students were actively seeking training in the Cepheid RT-PCR technology. One of CCBC's key problem solving technicians, Robert Shieman received a full scholarship to Stanford. Others like Dwylene Zapparelli, Steven Ekman, Thrace Allen, Maribel Aguilera, Nikki Geluz and Norma Villalon participated as a team in deploying the technology in workshops for industry employees at hotels and labs in the region. In one session for the Integrity Biosolutions manufacturing team, we were able to utilize Gram staining, RT-PCR, gel electrophoresis to confirm bacterial identities in 3 hours! Furthermore, Robert, Maribel, Steven and myself cloned and sequenced bacterial PCR fragments demonstrating proof of principle that this particular technology could also revolutionize the traditional curriculum in microbiology lab instruction. Now, Central Coast region colleges Ventura, Moorpark, College of the Canyons and Oxnard all include a detection module at some point during the year.

Convergent technologies like the Cepheid RT-PCR product involve combinations of biotechnology, nanotechnology, computer science and cognitive science. The cognitive science part is difficult to grasp, but is facilitated by software of an exceptionally good design. There will be many other technologies that are expected to follow this same development path. These technologies will in themselves create training needs for industry.

CONCLUSION

New technology deployment also roughly follows the Thomas Kuhn paradigm. The problem in microbiology is that from the top down, methodologies to determine bacterial species are standardized around biotechnical protocols that originated with Pasteur and Koch and were perfected in the first half of the 1900s. A new technology that incorporates previous capabilities of existing technology and resolves anomalies (the need for regulatory approval by FDA, EPA and DOD) is expected to generate some resistance by those supporting a simple and cheap (but slower) biochemical curriculum. The college faculty are ready to make the transition. There is a strong awareness that a true transformation of the diagnostic approach to keying bacteria with a combination of genetic and traditional biochemical methods is now within reach, because of the vast numbers of bacterial DNA sequences available in Genbank. DNA sequences can now be generated from primers used to amplify the 16S genes of any bacteria. A RT-PCR platform that has a set of protocols that are regulatory approved will promote its adoption in pathways leading to MLT (medical laboratory technician) or CLS (clinical laboratory specialist). The microbiology and biotechnology course infrastructure could make a significant contribution to training for those who could monitor food safety and water quality, check the health of hospital admissions, monitor bioreactor cultures and provide diagnostics at point of care. Many platforms are competing in this sector, but an implementation that is sturdy enough to also be used in training has been missing. The advantage to education is that the Cepheid unit actually contains 16 separate PCR machines—ideal for handling the asynchronously generated samples of students in a classroom. Furthermore, California is currently sending many healthcare samples out of state for determination of bacterial identity. The delays, increased healthcare costs, and degradation of patient care this entails are simply unacceptable.

California's biotechnology industry is expected to greatly simplify other complex technologies such as gas chromatography – mass spectroscopy (GC-MS), microarrays, DNA sequencing and lab-on-a-chip PCR. As instructors, economic development partners, industry consultants, and program developers, center directors must anticipate the simplification and convergence of emerging technologies so as to animate the region with new ideas for education.

The ability to popularize and institutionalize novel emerging and transformational technologies will determine California's success in competing globally. The work plan algorithm for developing and deploying these technologies is similar whether it is applied to federal, state, or foundation grants. The Central Coast Biotechnology Center participated in sparking adoption of several other emerging technologies. In 2003, a pilot-scale biodiesel refinery was built at the Port Hueneme Naval Base in conjunction with the CCBC, Biodiesel Industries, County of Ventura, National Park Service and the Federal Laboratory Consortium. A cellulosic biogas digester was completed by Onsite Power at U.C. Davis in 2007 following its incubation at CSUCI in the Gold Coast Innovation Center (CCBC served on the board). Currently, a suite of portable instruments is being examined for further regional deployment for a rapid bacterial detection training lab from Cepheid, Guava (liquid flow cytometry), and Climet (air particle sampler). Still other projects led to collaborations beyond the scope of the Center (see the *CSUCI MS Biotech and Bioinformatics Program* in the chapter titled *Establishment of a Successful PSM Degree in Biotechnology and Bioinformatics and an Innovative MS Biotechnology and MBA Dual Degree Program,* elsewhere in this volume). Dr. Harber is now focusing on developing the cluster defined by Oxnard Community College Marine Sciences Microbiology and Biotechnology, the rapidly expanding CSUCI campus, Camarillo and Thousand Oaks biotechnology businesses.

A strategy for developing a robust grant work plan is described here as an algorithm which involves the following steps:

1. Partnership with representatives of biotech industry, venture capital, economic development and education.
2. Formulation of informal and formal surveys to determine the evolving needs for skills training in the region.
3. Deployment of educational modules in regular courses and eventually in specialized courses. These same offerings are also offered for deployment in

the region directly to industry, venture or economic development representatives upon request.

4. Networking with the advisors in special advisory sessions that are part of larger industry specific meetings.

5. Identification of transformational, rapid and cost saving technologies that can be deployed within a single grant cycle.

6. Attraction of entry-level students to the emerging technology project to enable its deployment offsite and to the industry (because biotech involves labor for many pieces of equipment, even when it is simplified).

7. Deployment of the student run biotechnology teaching modules for industry representatives in workshops, forums and classrooms.

SPECIAL ACKNOWLEDGEMENTS:

Alzheimer's Institute: Gil Rishton

Amgen: Dr. Bruce Kerwin, Radu Georgescu, Milo Ferro, Mike Medlin, Yvonne Connell, Susanne Meyer, Jeff Lewis, and Dr. Timothy Osslund

Amgen Bruce Wallace Biotechnology Program: Hugh Nelson and Marty Ikkanda

Baxter Biosciences: Bill Tawil and Laurence Guiheen

Biodiesel Industries: Russ Teal

Bioenergy Producers Association: Dr. Kay Martin

Biolink: Jim DeKloe and Dr. Elaine Johnson

Bristol-Meyers Squibb: Ken Larson

Cal Poly San Luis Obispo and C3RP: Susan Opava

California Community Colleges: Jim Wolf, Mary Rees, Dr. Maureen Harrigan, Marie Panec, Marty Ikkanda and so many others.

Ceres: Ken Feldman and Richard Schneiberger

County of Ventura: Kathy Long

CSU Channel Islands: Dr. Richard Rush, Gary Berg, Marty de los Cobos, Dr. Ching-Hua Wang and Dr. Phil Hampton

CSU Channel Islands Business and Technology Partership: Gary Barsley, Wayne Davies, and Pauline Malysko

Dako: Uffe Lovborg

EWD Program: Jeffery O'Neil, Wendie Johnston, Sandra Slivka, Nora Lem, Edith Leonhardt, Ken Olson, Kay Ferrier, Terry Naylor and Teresa Parkison

Federal Technology Consortium: Kurt Buehler

Genetic Identification Services: Ken Jones

Gold Coast Innovation Center: Myron Miller, John Mungenast, Karol Pessin, and Craig Brown

Integrity Biosolutions: Byeong Chang and John Murray

Invenios/Girvan: Ken Richards

Invitrogen: Stuart Jaffe and Kevin Regan

National Park Service: Kent Bullard

Onsite Power: Orville Moe

SLO-EVC: Mike Manchak

SoCal Biotech Ventures: Brent Reinke

Tech Coast Angels: Gary Clark

Time Structures: Gus and Victoria Koehler

Ventura Community College: Dr. Robert Renger, William Thieman Barbara Hall, and Dr. Mary Pat Huxley, William Budke, Terry Pardee, Marta deJesus, and David Oliver

VCEDA: Bill Burratto, Darlene Ruz

Apologies in advance for any significant oversight of credits.
In memory of Bruce Wallace, who matched the CCBC grant from Amgen.

Biotechnology in Emerging Technologies:
A Case Study
Linnea Fletcher

Linnea Fletcher, Ph.D., is Chair of the Biotechnology Department at Austin Community College and Regional Director of Bio-link. Linnea can be contacted at *linneaf@austincc.edu.*

Many new technological advances are a result of a convergence between biotechnology, information technology, electronics, nanotechnology, and medical laboratory technology. Nanotechnology applications in biotechnology have opened new possibilities for examination and manipulation of matter at molecular and atomic levels. New techniques such as real-time PCR and microarray analysis have expanded the possibilities in diagnostic medicine. Medical lab testing processes, formerly conducted manually on large machines, have been automated using new equipment, including equipment in the nanoscale. There is a trend toward smaller equipment that can provide high throughput processing.

Maintaining this equipment requires bioinstrumentation specialists who have access to the modular boards that are part of the equipment. As a result, much of the repair work is outsourced to the site to where the equipment was manufactured, or at a site designated by the company; even overseas. Presently the companies providing this type of training start with individuals who have a basic knowledge of electronics and strong electromechanical skills. If any training is to be done by electronic programs in educational institutions, it will most likely be only ones near those companies. Since this equipment is usually very expensive, it would benefit educational institutions if there was also a biotechnology program, as the two programs could share equipment.

There is a large foreseeable need for medical technologists (MLT) who can operate the new diagnostic equipment, and understand the biotechnological basis for the assays performed. Many existing technicians operate equipment without this knowledge and, as a result, find it difficult to troubleshoot assay problems. MLT programs would benefit from partnerships with biotechnology training programs as many of them already provide training on this equipment.

Obviously, departments within educational institutions should be encouraged to work together to meet the needs of this industry and their students. Traditionally this has not always been the case and departments are often indirectly encouraged to compete against each other for institutional and outside funding. Interdepartmental competition leads to duplication of efforts and equipment which increases institutional spending, decreasing the institution's ability to effectively meet the needs of industry. Industry would be better served if departments were financially encouraged to work together sharing resources including information, faculty, supplies and equipment. This chapter reviews what three departments at Austin Community College (ACC) are doing together as a team to meet these needs.

In August 2006, the Austin Biotech Workforce Education Consortium was awarded a state grant beginning a two year project to expand and enhance biotechnical educational offerings at ACC based on the emerging technology needs of medical laboratories, and bioscience companies in the areas of bioinstrumentation and medical diagnostics This consortium consists of *Work*Source – Greater Austin Area Workforce Board and its key stakeholder partners including the electronics, biotechnology, and medical laboratory departments at ACC, The Greater Austin Chamber of Commerce (GACC), and area employers. The external evaluator was Bob Glover of the University of Texas' Ray Marshall Center (RMC). RMC conducts applied interdisciplinary research to foster improved policies, programs and practices in education, workforce development, child care, welfare, and social policy. Together, the departments are determining new skill and knowledge competencies, developing certificates and degrees incorporating these competencies. In addition, plans are being implemented that share space, faculty, staff, supplies, and equipment so that, after the grant ends, they can continue to work together.

A program's prior knowledge and actions influences future approaches. Before providing information of what the departments are doing together, it is important to understand how each one is both physically and philosophically structured to provide the best training for its students. Just like students, faculty's (i.e. departments) prior knowledge and actions affect how they internalize new information and perceive educational problems, and therefore how they will want to develop curriculum and programs. If departments and outside partners want to work together, it is important that each one understands and appreciates the other's approach to solving educational problems.

A description of each department, starting with the biotechnology program, follows.

THE BIOTECHNOLOGY PROGRAM

Many roles in the biotechnology industry do not require employees to gain national or state certificates. As such, educational programs have a greater leeway on how they decide to meet the needs of industry in their area. The ACC Biotechnology Program offers a variety of educational pathways for students. For 2-year students, it offers a 1-year Certificate in Biomanufacturing, an AAS in Biotechnology, and for students who already have a 4-year degree in an appropriate field such as biology, biochemistry or molecular biology, a 1-year Advanced Technical Certificate. The courses are also offered as continuing education courses for individuals looking to expand their skill sets without getting a certificate or a degree. Industry requested that the courses be offered at night so that students could be hired before finishing their degree and so that employees could take courses. There are a total of 7 biotechnology courses; the first year focuses on basic skills such as micropipetting, preparing solutions, and performing assays, while the second year focuses on more specialized skills in cell culture, DNA/RNA techniques, and protein purification. Skills and knowledge built into the program include safety training, good manufacturing practices, FDA regulations, ethics, and teamwork. Courses are mainly taught by industry employees; this ensures that courses are current and meet industry needs. Over time, faculty have come to realize that the program needed to be run more like a company in order to better train students for jobs in the industry. As a result, students are evaluated in every course using a personal evaluation adapted from local industry, and they are required to pass both skill and knowledge tests before being allowed to progress to the next level. Class size limits are kept low to ensure that students are given individual attention and, when necessary, can perform job functions individually to ensure they have mastered techniques. To progress to the final course, internship, students are required to pass a faculty interview to ensure they are ready for the interviews they will face when applying for company internships. After they finish their course of study, the placement rate for students is 95 percent. One key factor in strong student placement is the individualized attention ACC's biotechnology department provides to in a "case management" style. Each student receives counseling on his/her degree plan and employment plans.

HIGH SCHOOL PROGRAM

With funding from an National Science Foundation (NSF) Advance Technology Education (ATE) grant several years ago, a high school biotechnology program was established. It was supposed to establish a major pipeline of students into the 2 year program, but that has not been the case. However, the program has expanded into more high schools and, based on the qualifications of the teacher, students receive dual credit (high school and college) for the one year biotechnology course. It is equivalent to the one semester introductory biotechnology course offered in the two year program. The program trains teachers and provides equipment and consumable supplies. As a result of this experience, a high percentage of these students continue their education in a biotechnology-related field at 4-year schools; they also get hired by industry right out of high school based on the entry-level skill set they obtain through the course. Some of the students come back to ACC to obtain the Advanced Technical Certificate once they finish their 4-year degree. Most students do not directly enter the two year program, as we have found that students are still being encouraged to attend 4-year schools by parents, counselors, and their own expectations for success; this may change with time.

ELECTRONICS AND ADVANCED TECHNOLOGIES

Austin is a major semiconductor hub, and as such the ACC electronics department is well funded and has state of the art equipment, including a clean room. The existing department chair was recruited from the semiconductor industry and has re-done the degrees and certificates such that there is an establish core set of courses all students finish before going on to a special-ization such as robotics or semiconductors. Thus students receive an AAS with a specialization, and there are seven specializations that students can receive. This department also offers both credit and continuing education courses, and also has several courses in area high schools. Just like the Biotechnology Program, class siz-es are small so that individual attention can be given to students and also ensure that all students have mastered certain techniques. The department chair did an exhaustive analysis of student suc-cess rates in core classes, and upon discovering which ones had the highest failure rates, increased the number of available tutors for those courses. This has decreased the failure rate for those courses. Courses are offered both day and night. The department chair is also very successful in getting grants at both the state and national level, so there is constant adjustment of the curriculum and pro-grams. Like the Biotechnology Program, this program has a major presence in area high schools.

MEDICAL TECHNOLOGY (MLT) PROGRAM

In Texas, clinical laboratories can hire both 2-year and 4-year degree students with clinical laboratory science degrees, even though the degrees are different. ACC offers a two year Associate Degree in Medical Laboratory Technology. The program is accredited by the National Association for Clinical Laboratory Sciences. After students finish their degree, whether it is 2-year or 4-year degree, they must pass a national certification exam before they are allowed to work in a clinical laboratory. This means that, unlike the Biotechnology Program, the MLT program has a more nationally standardized curriculum and must constantly ensure that their program, including faculty, maintains national standards to remain accredited. The MLT program incorporates both credit and continuing education courses online, in the classroom, and in the laboratory training for their students. The majority of classes are held during the day.

As indicated above, the MLT and Biotechnology Programs share the same teaching laboratory space. Their preparation rooms are separate and the biotechnology cell culture room is located in the biotechnology prep room. The two programs have separate laboratory assistants and are supervised by different deans. The Biotechnology Program is supervised by the dean of Math and Sciences while the MLT Program is supervised by the Allied Health Science dean. The Biotechnology Program has benefited from this sharing of space in that it has raised the level of its safety training in line with MLT requirements. The MLT program has benefited from the Biotechnology Program expertise in molecular biology techniques as well as from access to molecular biology equipment.

EVENTS LEADING TO THE GRANT

Prior to awarding of the grant, the ACC MLT Program had already discovered that their graduates and incumbent workers were going to need training in molecular diagnostics as the growth in the area of molecular based testing is growing at a phenomenal rate. The Biotechnology Program saw that many of the companies hiring their students were focused on development and production of molecular diagnostic products. However, even though both programs realized these changes, neither program had the time to collaborate on an obvious shared goal of training students in molecular diagnostics. Release time was needed for key people in both programs, and so the area workforce board was approached with the idea of obtaining a state grant that would fund release time and curriculum development. At the same time, based on collaboration on a Nanotechnology State Grant, the Biotechnology Program and the Electronics Program discovered that several bioscience companies in the area could benefit from employees with expertise in developing and fixing bioinstrumentation equipment. Biodevice companies had approached the Electronic Department Chair about hiring electronic graduates with bioinstrumentation expertise, and a bioscience company had hired a biotechnology graduate who was formally a semiconductor employee. It was decided to seek funding (release time for an electronic faculty member and for curriculum development) on this same grant for development of a bioinstrumentation specialization within the electronics department.

GRANT WORK

The department chairs of the Biotechnology and MLT Programs received full release time for the first year of the grant to visit clinical laboratories and biotechnology companies using or producing molecular diagnostic kits or approaches. It was discovered that many of the technicians in the clinical laboratories did not understand the molecular basis for their tests. It was also discovered that several of the biotechnology companies making these kits would consider hiring MLT students if they understood the molecular biology behind these kits and improved their basic laboratory skills such as micropipetting. The MLT department chair participated in a certification program in molecular diagnostics offered by the Molecular Laboratory Diagnostics Program at Michigan State University[1]. The program involved taking two classes online combined with one week of intensive hands-on clinical wet lab experience at the Michigan State campus in Lansing. The biotechnology department chair soon realized that it would be advantageous for her to obtain the same certification as it was obvious that the MLT department chair was going to base a lot of the developed curriculum on that program. Even though the biotechnology department chair knows quite a lot about the techniques and the molecular biology behind the diagnostic testing, she found it worthwhile to learn about the applications and also be "on the same page" as the MLT department chair when developing curriculum. Based on the needs of industry, incumbent workers, and students, it was decided that a one-year specialty certificate would be appropriate. The prerequisite for the certificate is either a 2-year or 4-year degree in clinical laboratory science. The certificate program is described at *www.austincc.edu/mlt/md/mdcertificate.html*. The first course is to be taught online, just like the first course in the Michigan State program. The second course will be a combination of online and in the laboratory training and will be taught by a biotechnology faculty member. The final course is an internship at an appropriate site. If a student is already an employee of a laboratory doing diagnostic testing, the position would serve as the internship.

The three courses are:

MLAB 2378 *Fundamentals of Molecular Diagnostics*	*4 college credits*
MLAB 2479 *Molecular Diagnostics Techniques*	*4 college credits*
MLAB 2363 *Molecular Diagnostics Clinical*	*1 college credit*

When employees of existing laboratories found out that the program would start fall 2007 and that tuition was being paid for by the grant, they immediately signed up. Industry appreciated the immediate and tangible results they obtained after all of the time and trouble they spent in providing faculty externships or assisting the project.

---

1   *http://bld.msu.edu/mol_lab.html*

The Electronics and Advanced Technologies Department chair was too busy to work on this project so he drafted a full-time faculty member who received a partial course load reduction. Not being part of the initial team that conceptualized the grant, this faculty member initially had a difficult time recruiting additional companies to participate on the grant. Also, the companies that had provided the initial positive feedback for the grant were not available to work with, as they had been sold or were being reorganized, and the original contact people had left the companies. Luckily additional companies were located with the help of the MLT department chair and they were willing to donate equipment and suggestions. The faculty member commented that the equipment has similar components to other electronic equipment that he used and taught students how to use. This faculty member was also willing to attend a national biotechnology meeting and meet with other faculty across the nation and to visit a major biotechnology instrumentation company to develop his own contacts within this industry. Since the courses needed to fit the existing framework for this department (they were core courses with a specialization), but they also needed to fit the time frame of the grant, it was decided to develop both a specialization for students taking the electronics core and also a 1-year advanced certification for employees in a related electronic industry to add a bioinstrumentation specialty. One of the courses in the specialization is *BITC 1200 Fundamentals of Biotechnology*, also part of the biotechnology program. It was felt that it would be easier to have a biotechnology course as part of the specialization.

The courses being developed are:

INTC 1491 *Biotechnological Instrumentation*	*4 credit hours*
INTC 1491 *Biomedical Instrumentation*	*4 credit hours*
EECT 2188 *Internship*	*1 credit hour*

Fifteen students were initially recruited for the first class in fall 2007. About half are incumbent workers and half are existing students in the Electronics and Advanced Technologies Department.

The grant is paying the tuition for the first group of students taking these three courses.

FUTURE PLANS

The grant allowed faculty from different departments sufficient time to work together, visit companies, and meet employers. They were able to develop their own relationships with employers and employees in the other fields which is important for the sustainability of their new certificates and degrees and for meeting the future needs of the industry.

Other collaborative possibilities being developed, not part of the original grant, are:

- Sharing space (e.g. the clean room) and new equipment
- Development of a donated and used equipment depot that is shared by several departments.
- Development of shared initiatives between different areas of the college
- Sharing faculty and courses

Busy, goaled-oriented individuals will always find it hard to find time to meet and share ideas and goals. It is hoped that this collaboration will continue, but it will take extra effort on the part of all parties as all of the individuals, departments, and partners are involved in so many projects that collaborative projects, requiring them to consider other points of view, etc., tend to receive low priority. It is suggested that educational institutions routinely require departments from different areas of the college meet and/or provide more funding opportunities requiring a sharing of resources. Industry realizes that this it is important to expand and form partnerships outside of their original area of expertise, and so should educational institutions and the departments within those institutions.

The Professional Master of Bioscience Program at Keck Graduate Institute

Jane G. Tirrell and T. Gregory Dewey

Jane G. Tirrell, Ph.D., is Assistant Dean of Faculty at Keck Graduate Institute of Applied Life Sciences in Claremont, CA. Jane can be contacted at *jane_tirrell@kgi.edu.*

T. Gregory Dewey, Ph.D., is Senior Vice President for Academic Affairs, and Dean of Faculty at Keck Graduate Institute of Applied Life Sciences in Claremont, CA. Greg can be contacted at *greg_dewey@kgi.edu.*

A number of high profile publications[1,2,3] have been critical of the nature of doctoral education in the United States. The common criticism is that PhD training takes too long, is too narrowly focused, and does not confer the skill sets required in the market place. At a time when there is a glut of PhD graduates across the sciences, there are unmet needs in industry for technically competent individuals who are not practicing scientists.

The marketplace is becoming increasingly technology-driven and there is a strong need for individuals who have the quantitative skills exhibited by PhDs and the management skills of MBAs. The bioscience industry provides a particularly striking example of the need for technologically competent managers; the genomic revolution is having an enormous impact in directing the biotech, pharmaceutical, and medical device industries into increasingly

1 *Committee on Science, Engineering and Public Policy (1995) Reshaping the Graduate Education of Scientists and Engineers, National Academy Press, Washington, D.C.*

2 *Committee on Graduate Education (1998) Report and Recommendations, American Association of Universities, Washington, D.C.*

3 *Golde, C.M. & Dore, T.M. (2001) At Cross Purposes: What the experiences of doctoral students reveal about doctoral education (www.phd-survey. org). Philadelphia, PA: A report prepared for The Pew Charitable Trusts.*

high tech domains. The dominant role of regulatory agencies in industry activities gives rise to a critical need for technically trained individuals working in non–technical positions.

To meet these workforce needs, Keck Graduate Institute (KGI) has established the Master of Bioscience program. Nationally, an assortment of professional master's programs is arising in a variety of institutional settings. These programs take diverse forms but most offer technical courses of study in addition to a management component or industry internship. These programs span a range of areas including applied physics, mathematical finance, biotech, and computational biology. Their goals are often quite broad in conjunction with their intentions of offering alternatives to traditional PhDs and MBAs. The effectiveness of these programs will, to a large extent, be dictated by the institutions' abilities to establish them as recognized, respected degree options. In most pure sciences, the Master of Science degree does not garner much respect and is frequently viewed as a "failed PhD." For professional master's programs to establish themselves, these attitudes must be dispelled. However, similar to the MBA, the professional master's degree has defined a niche in our technology-driven society. Given the strong educational need for technically sophisticated individuals, the trend toward professional master's degrees is an inevitable response to market forces.

THE NEED FOR NEW TRAINING IN THE BIOLOGICAL SCIENCES

The Human Genome Project has advanced change in the biological sciences. This project integrated molecular biology, laboratory automation and miniaturization, and computer science to yield a database of human information of indisputable importance for the medical, pharmaceutical and health care industries. While it was a tour de force of laboratory automation technology, the project also demonstrated the power and impact of computation. High power computation, in combination with the high throughput technology of gene sequencing, allowed the project to confound the critics and be successfully completed well ahead of schedule. The project was touted as a harbinger of the future importance of biology and the applied life sciences.

In the current era, sometimes referred to as the "post genomic era," high throughput technologies are proliferating that allow the analysis of the molecular constituents of a living organism to occur with unprecedented accuracy and scope. Genome-wide technologies enable the monitoring in a single experiment of the expression of every gene in a living organism. "Lab on a chip" devices make possible rapid and very high throughput diagnostics for examining biological responses to toxins, drugs, diseases and other environmental stresses. These technologies in conjunction with extensive computational resources are yielding new fields such as structural genomics, functional genomics, pharmacogenomics and proteomics. The need for computation, integration, and interpretation of biological information is fueling the new disciplines of computational biology and bioinformatics. The dramatic changes of this era are not only altering how biology is done, but the very nature of biology itself. Biologists are now asking very different questions and are beginning to tackle the system-wide complexity existing at the molecular level of basic life processes.

As the practice of biology changes, so must the education of biological scientists. The challenge for KGI's MBS program has been the lack of "road maps" for effecting these changes. There are no established curricula, no textbooks and no well-defined disciplinary boundaries. In lieu of such guidelines, we operated from a series of basic premises in constructing the MBS curriculum. We briefly examine these foundational premises.

Biological sciences have been transformed by computational and high throughput technologies; these changes will also impact the applied life sciences and how they are done.

The transformation of the biological sciences offers exciting challenges for the training of future biologists. In this new era, biologists must be educated in a much more quantitative fashion than in the past. This shift in education must occur at all levels as has been noted in the recent study, *Bio 2010: Undergraduate Education to Prepare Biomedical Research Scientists*[4]. This study calls for a wholesale revamping of the undergraduate curriculum in the biological sciences for more effective training of doctors and medical scientists. Bio 2010 calls for much more quantitative teaching of biology and a more thorough integration of topics from chemistry, physics, and engineering into the biology curriculum. This study reinforces our experiences at KGI. The fundamental changes in biology will have far-reaching effects that are being transmitted throughout the life science industry. For the pharmaceutical and biotech industries, this is evidenced by changes in the drug discovery process, the high interest in genotyping technologies and the impact of pharmacogenomics. In the medical diagnostics industry, there is a drive toward molecular sensors and diagnostics that reflects these fundamental changes in biology. The medical device industry is utilizing a deeper understanding of biocompatibility to create a new generation of implantable devices.

While the technical expertise required by industry cannot be fully anticipated, computational biology, bioengineering and systems biology will be the underlying scientific disciplines that support discovery and development.

This premise was spurred by the philosophy that it is better to prepare people for future changes than for present jobs. Thus, the curriculum is based on fundamental knowledge from the three disciplines mentioned above. The level of integration of the science and engineering curriculum is a key feature that anticipates the more quantitative future nature of the applied life sciences. This integration can be accomplished through applications-based projects that allow interplay of the three technical areas. When incorporated into the curriculum, such projects allow material to be presented much as it would appear in a real-world setting. There is an inevitable tension between training that supplies specific skills for specific jobs versus a deeper more adaptable education for op-

---

4   *Committee on Undergraduate Biology Education to Prepare Research Scientists for the 21st Century (2002) Bio 2010: Undergraduate Education to Prepare Biomedical Research Scientists, National Academy Press, Washington, D.C.*

erating in a shifting industrial environment.

While the professional skills required by industry cannot be fully anticipated, individuals will be required to work in interdisciplinary teams on applied problems, to communicate well, to have management capabilities and to understand ethical issues.

As the life sciences industry responds to the underlying changes in biological sciences, we will see an increasing emphasis on group work involving individuals from multiple disciplines. Applied problems are most effectively attacked by groups having a range of expertise that also may include management and ethics components. Our students must learn to work in and manage groups, and be able to communicate with people of diverse backgrounds. This dynamic requires that students have a basic knowledge of business organization, finance, strategy, marketing, and project management. They must also be able to identify ethical issues and have a framework for dealing with them.

The best managers will be scientifically trained. Scientists and engineers with management training will perform better than technical MBAs.

This premise goes to the core of the curriculum and the mission of the Master of Bioscience program and motivates the structure of KGI's entire program. The program integrates science and engineering with management and ethics. The basis of this premise is the conviction that the future of applied life sciences will involve such a sophisticated interplay of disciplines, that management training must be coupled to a strong technically based education. This premise also supports the notion of generalized education across disciplines rather than highly specialized training. Generalists who communicate well across disciplines will be better managers than people who are overly specialized in a single discipline.

STRUCTURE OF THE CURRICULUM AND THE KGI EXPERIENCE

In this section, we provide a synopsis of our experiences to date with the Master of Bioscience curriculum and associated pedagogy. We state the motivation for each curricular element and describe how these elements fit into the overall program.

The curriculum contains the following elements:
- A short *initial project* that serves as an immersion into team projects and the life science industry.
- A set of *first-year courses* that give the student a broad understanding of the life sciences industry.
- A summer, paid *internship* in the life sciences industry that provides practical workplace experience.
- A set of second year courses that form a *focus track* that allows students to specialize in specific career-oriented areas.
- A *Team Masters Project (TMP)* in which student teams work on industry sponsored projects as a capstone experience.

These curricular features along with their learning goals are described below.

INITIAL PROJECT

Upon arrival at KGI, the first-year class spends ten days working in teams of three to five students each on the Initial Projects. The *Initial Project* is an assignment to perform a retrospective analysis of some event that occurred in the applied life sciences industry, for example, the discovery of a new drug or the development of a new technology or device.

Titles of past *Initial Projects* include:
- *HIV Therapeutics & Agouron Pharmaceuticals:*

Development of a Protease Inhibitor Drug

- *Genentech and the Development of Human Growth Hormone*
- *The Development and Marketing of Insect-Resistant Bt Crops*
- *Advanced Tissue Sciences, a Pioneer in Tissue Engineering*
- *Aquagene: Transgenic Fish as Bioreactors for Complex Human Therapeutic Proteins*

The student teams analyze the event from a range of perspectives, considering the scientific/technical issues, the intellectual property issues, the marketing and business aspects of the development and any associated ethical issues. They analyze a company's performance with the benefit of hindsight and critique decisions made during product development. For each *Initial Project*, an expert who had been involved in the event, often the CEO of the company concerned, is available to the student team to discuss the company's decisions. During this ten–day period students also attend workshops on team building and communications skills. The *Initial Project* concludes with a day of public presentations and the submission of a written report by each team. The *Initial Project* is not graded, but a prize is given for the best presentation.

The learning goals of the *Initial Project* are to give the students knowledge of the business practices of life science industries, to teach them to work in interdisciplinary teams, to address complex problems, to develop writing and oral presentation skills and to develop an awareness of how ethical issues creep into industrial problems. From an assessment point of view, this project serves as a "book-end", revealing the students' level of knowledge as they enter the program. The other book-end consists of a similar assessment of *Team Masters Projects* completed just before the students graduate.

The *Initial Project* is a total immersion in the affairs of the bioscience industry. Students are exposed to issues previously unfamiliar to them and are introduced to practices such as patenting and intellectual property protection that they rarely, if ever, encounter as undergraduates. They see real world examples that show that a company needs to do more than just get the science right. The students are required to work in teams, manage their time, and communicate orally and in writing. The exercise generates an excitement about the MBS program among the students and provides a rapid and effective introduction to the life science industry. Students become intensely engaged in this exercise.

FIRST-YEAR CURRICULUM

Our students come from diverse academic backgrounds. As shown in Figure 1, 50 to 60 percent have undergraduate majors in biological sciences and the remainder have majors in engineering, the physical sciences, and other disciplines that include math and computer science.

They come from a wide assortment of schools and include a significant international student population (20-30 percent). Because of the breadth of the curriculum and the academic diversity of the class, we find it necessary to integrate background–filling material into the common set of first-year courses. We cannot expect that this material will completely "level the playing field," rather it is intended to introduce the students to the basic concepts and language necessary for continuing in the MBS program. We strive to enable the students to identify and frame problems in disciplines outside their undergraduate education, to communicate with people from fields outside of their expertise, and to know where to find the resources to solve problems.

The first-year curriculum at KGI is a fixed set of courses required of all students that incorporates science and engineering, management, pharmaceutical development and bioethics. In addition to the courses providing background-filling (*Computational and Mathematical Methods* and *Molecular Biotechnology*), the first year contains the technical survey courses, *Systems Biology of Complex Diseases, Computational Biology, Medical Devices and Diagnostics, Bioprocessing, Biologics and Pharmaceutical Discovery and Development.* These courses cover the industry and offer the basic science con-

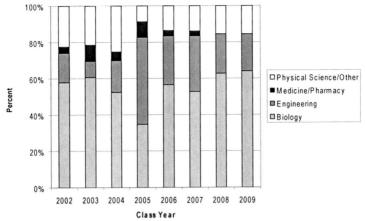

Figure 1: Diverse academic backgrounds of KGI students

tent of the program. The management courses in the first year include *Accounting, Finance, Organizational Behavior,* and *Strategy and Marketing.* The challenge of the first year has been to integrate topics not only within the sciences but between the sciences and business. Through informal oral feedback and course evaluations, the faculty realize that students feel that they are faced with a bewildering array of challenging topics and do not always see how these topics fit together. However, students often come to appreciate retrospectively the relationships between the different areas. Usually, by the middle of the second semester, students begin to see the "big picture." Courses such as *Pharmaceutical Development* play an increasingly important role as bridges between the technical and management courses.

The learning goals for the first year are broad, encompassing virtually all of the goals in the assessment plan for the **MBS** program (see Table 1).

The first year is directed to the students' acquiring general knowledge, while the second year deals with more career-oriented education. As in all aspects of the curriculum, the learning goals are divided into three main areas: knowledge, values, and professional skills. The learning goals associated with professional skills and with values are common to both years of the program. Numerous first-year projects provide opportunities for program assessment. Of par-

Table 1: Alignment of learning outcomes with curricular milestones

Learning Outcomes / Curricular Milestones	Student CVs	Initial Project Oral Presentations	Initial Project Written Reports	Medical Diagnostics Team Project Report	Marketing Analysis	Bioethics Oral and Written Reports	Industry Internship Poster Presentation	Industry Internship Written Report	Team Masters Projects Oral Presentation	Applied Entrepreneurship Business Plan	Completion of Career-Oriented Focus Area Requirements	Exit Interviews
MBS students can integrate the fundamentals of computational and informational science, engineering design, and biomolecular technologies to solve problems in applied life science.		▓	▓	▓				▓	▓	▓		
MBS students can contribute productively on an interdisciplinary team tackling complex problems.		▓	▓	▓					▓	▓		
MBS students can communicate effectively in an industry environment composed of scientists, engineers and business professionals.	▓	▓	▓	▓		▓	▓	▓	▓			▓
MBS students have the core business analysis and management knowledge needed for the bioscience industry and can assume leadership roles in realizing the goals of technical and business projects.				▓	▓				▓	▓		
MBS students understand the business, research and development, regulatory, production, and marketing functions of the bioscience industry.				▓	▓		▓	▓		▓		
MBS students have both broad understanding of systems biology, bioengineering and computational biology, and specialized knowledge required for a specific career track in the applied life sciences industries.										▓	▓	
MBS students value the translation of basic science and engineering discoveries into products and processes, which benefit society.	▓										▓	▓
MBS students adhere to ethical principles in research, development and business issues inherent in the bioscience industries.						▓			▓			▓

ticular interest is the final design project in the *Medical Diagnostics* class. This project involves role-playing where teams of students assume an organizational structure of a company with individuals assigned to different roles within the company. In this assignment, faculty assess team-learning on inquiry-based projects. The assessment rubric shown in Table 2 contains faculty comments on performance of a team involved in designing a drug-eluting stent.

INTERNSHIP

During the summer following the first year of courses, the students participate in a paid internship in a bioscience company. The summer internship engages students in the current practices of a particular industrial sector and helps develop an understanding of the climate and culture of business. The learning goals include development of professional skills associated with the specific area of the internship, as well as an understanding of the corporate values of the sponsoring company. In addition to fulfilling the job responsibilities of the internship, the students observe and report on the organizational structure of the sponsor company. At the end of the summer the students give public poster presentations on their internship experiences.

Table 2: Assessment rubric

Objective:
Students know how to organize within a team and develop roles suited to particular tasks.

Performance Criteria	Exceptional	Emerging	Below Expectations
Clear missions are defined for the teams	Statements of each team's purpose outlined in report Executive Summary.		
Tasks and responsibilities are clearly identified	Work breakdown structure chart shown for each subsystem team.		
Team meetings are effective	Comments from peer assessments of team members' performance indicate that good progress was made in team meetings.		

Student feedback, gleaned primarily through exit interviews, confirms that the internships are a positive experience that helps put the MBS curriculum in context. In general the sponsors have been extremely satisfied with the quality of the students, and internships have frequently resulted in job offers. Several of the companies have also become more engaged with KGI as demonstrated through their sponsorship of *Team Masters Projects (TMPs)* and involvement on our advisory council.

SECOND-YEAR FOCUS AREAS

In the second year, students take courses in career-oriented focus areas and have an opportunity to achieve some depth in an area of interest. The course offerings have evolved from a collection of in-depth graduate courses to a set of specific tracks that guide the student to a sequence of courses appropriate for his or her career interests. These courses are taught in a variety of formats: from traditional lecture style to intensely lab-oriented classes. The current focus areas are:

1. Pharmaceutical Discovery and Development
2. Business of Bioscience
3. Clinical and Regulatory Affairs
4. Bioprocessing
5. Medical Devices and Diagnostics

Course requirements for the focus tracks consist of a sequence of four semester courses, augmented with business or science electives and with the TMP. These second-year offerings more directly match career tracks than the general education of the first year. An example for the Clinical and Regulatory Affairs track is shown in Table 3.

The second-year courses have been well received by the students. The courses contain considerable foundational material even though they are focused on career-related information. There is a persistent tension in the curriculum between delivering specific job-related skills versus more depth in a given field. The MBS curriculum does not train technicians nor are students requesting training of this type. Our presumption is that specific career areas are best studied in the context of project-oriented activities involving a range of disciplines. This presumption will be tested by the experiences of our alumni in time.

TEAM MASTERS PROJECT (TMP)

The team masters project is designed as the capstone experience. It replaces the traditional master's thesis project of conventional MS programs. The TMP has corporate sponsorship and very specific goals and deliverables. A description of a recent TMP is given in Figure 2. The team views the corporate sponsor as its client and the industrial problem to be tackled by the student team

Table 3: Required courses and electives for the clinical and regulatory affairs focus area

Clinical and Regulatory Affairs	
Fall	Spring
Technical Courses (two or more units per semester): ALS 435 US Regulatory Affairs ALS 437 Introduction to Clinical Pharmacology I ALS 434 Clinical Biostatistics ALS 401 Biopharma: Biotechnology-based Therapeutics **Electives (at least one unit per semester):** ALS 453 Biotechnology Law and Regulation ALS 455 Bioscience Business Strategy ALS 457 Building an Entreprenuerial Organization ALS 458 Applied Entrepreneurship: Writing a Bioscience Business Plan ALS 459 Managing Projects: A Dynamic View	**Technical Courses (two or more units per semester):** ALS 436 Adv. Pharmaceutical Development: Case Studies ALS 438 Introduction to Clinical Pharmacology II ALS 433 Clinical Trials Design, Conduct and Strategy ALS 426 Medical Device Development and Market Release **Electives (at least one unit per semester):** ALS 451 Biotechnology Intellectual Property and Licensing ALS 452 Entrepreneurship Practicum ALS 458 Applied Entrepreneurship: Writing a Bioscience Business Plan

is one for which the client has need of a solution or answer; that is, projects have a real potential payoff for the client. Further, KGI particularly seeks projects that involve multiple disciplines, including, if possible, both science/technology and management/ethics issues, thus mirroring KGI's overall curriculum. Ideally, the project involves the application of science/technology rather than the pursuit of new discoveries, and is more complex and challenging than simple routine testing or computer programming.

The TMP emphasizes problem-solving, project management, budget management, productive teamwork, and effective communication—skills that are critical to KGI graduates' careers in bioscience. Student teams typically consist of three or four second-year KGI students and are advised by a KGI faculty member. In the last two years, first-year students have been permitted to participate in TMPs in their second semester. The teams meet at least bi-weekly with the faculty advisors, and have contact with the company liaisons as needed, but at least bi-weekly. Each team provides its sponsoring company with an interim report at the end of the first semester and a final report, in lieu of a master's thesis, is submitted at the end of the year. All teams give public presentations of their projects at the end of the academic year. Teams are also available to make presentations of their work at the sponsoring companies' sites.

As mentioned above, assessment of student performance in the *Initial Projects* soon after they arrive at KGI provides a book-end for the extent of their knowledge and skills at the beginning of the program. Similarly, assessment of TMP performance provides the second book-end as the students are completing the program. Since the ability to work in teams and communicate effectively are strongly emphasized in the MBS program, we compared students' presentation skills for the Initial Projects and the TMPs using the rubric shown in Table 4.

We observe a slight, but significant improvement for nearly all of the nine performance criteria listed in the rubric. Figure 3 shows median rubric scores for *Initial Project* presentations (diamonds) and TMP presentations (squares). A Wilcoxon rank sum

Diagnostic Enzymes

BioCatalytics produces a comprehensive array of enzymes useful in the synthesis of new drugs. These enzymes may also have applications as diagnostics in food and environmental safety testing, clinical diagnostics, and research. In 2004, BioCatalytics created a small group of diagnostic enzyme products for the purpose of evaluating the sales potential for these applications.

The KGI team conducted a market segmentation analysis to assess potential customer demands and current competitors in the diagnostic enzyme arena. The project also included assay development and construction of manufacturing protocols for a few identified products. The team analyzed regulatory and manufacturing issues associated with launching diagnostic enzyme products.

Figure 2: Description of BioCatalytics Team Masters Project

test was conducted to determine p values; all p values except that for the Strength of Argument criterion, for which $p = 0.49$, were less than or equal to 0.02. All projects conducted by students in the classes of 2005, 2006 and 2007 are included in these results. Figure 3 indicates that students need some guidance in the proper citation of information sources.

Additional assessment of the effectiveness of the TMPs is obtained from surveys and interviews of the corporate sponsors and KGI students and faculty. After nearly seven years of experience with TMPs, we note that the most successful projects, as judged by accomplishing goals and meeting timelines, have had the guidance of conscientious and engaged industrial liaisons. Projects requiring laboratory work can be unpredictable with respect to delivering results. Faculty perceptions are that projects where tasks can be pursued in parallel tend to proceed more quickly than those for which the tasks are sequential. The students' responses to the projects have been variable. Most students struggle with the teamwork, but almost all students believe that participating in a team made up of people from diverse backgrounds and who have different work styles is a valuable learning experience. Some teams have felt the faculty should be more actively involved; however, the faculty advisor's role is only to provide guidance; they are not meant to direct the projects.

Table 4: Presentation rubric

Learning Objective:
Students can organize effective formal presentations of technical and business information to diverse audiences.

	Exceptional	3	Emerging	2	Below Expectations	1
Problem Statement	Clear, concise, accessible to broad audience		Generally focused and technically correct but takes some effort to understand.		Problem never precisely defined, unfocused and rambling or incorrect	
Organization	Clear, logical flow to presentation; transparent structure/outline		Mostly logical flow to presentation, placement of some information unclear		No clear structure or flow to presentation	
Depiction of information	Attractive format, balanced use of text, graphics and images; data clearly interpretable		Some use of graphics and images; data generally easy to interpret		Predominantly text; data poorly formatted	
Slide format and number	Each point is clear and succinct in appropriate font; right number of slides/time slot		Most slides clear and succinct with appropriate font; roughly appropriate number for time slot		Too much text/slide or points too vague to be intelligible; far too many or too few slides/timeslot	
Content	Content is accurate and complete		Content is generally accurate but not entirely complete		Content is inaccurate, overly general, and incomplete	
Strength of argument	Data provide convincing evidence to support conclusions		Data provide acceptable but not overly strong support for conclusions		Conclusions are not supported by sufficient evidence	
Originality	Provides significant new insights about the topic		Provides some interesting facts but limited insights		Audience is unlikely to learn anything or may be misled.	
Citation of information sources	Properly formatted citations are inserted wherever necessary		Citations in most necessary places		Inadequate citations or no citations at all	
Technical and business information	Captures key technical developments affecting the sector and explains their business implications to informed layman		Captures most technical developments affecting the sector and explains their business implications, but relies on arcane language		Fails to capture key technical developments affecting the sector; business implications poorly explained	

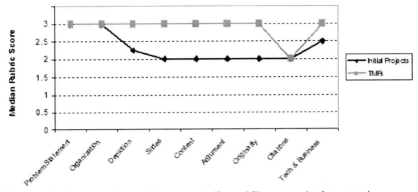

Figure 3: Comparison of presentation skills on entering and leaving MBS program

Both faculty and students agree that the projects chosen for TMPs need to be carefully evaluated. We have noted that projects involving technologies in a very early stage of development can be extremely unpredictable in yielding useful results. These projects have often not provided satisfactory experiences for the students.

EMPLOYMENT OF KGI STUDENTS

The ultimate proof of the effectiveness of KGI's professional Master of Bioscience degree program lies in the employability of our graduates. Through the MBS program, we have developed increasingly extensive networks within the life sciences industry that greatly assist in graduate employment. In fact, the percentage of the graduating class that has accepted a job or received an offer by the December before graduation increased from 3% in 2005 to 35% in 2007. Figure 4 shows graduates' first positions by job function.

Surveys of the MBS alumni conducted two years after graduation indicate that acquiring the ability to work in teams is one of the most valuable experiences of the KGI education. Alumni also mention that KGI's focus on curriculum at the interface of technology and business has been very helpful in their careers and they recommend that KGI increase its course offerings at this interface.

CHALLENGES AND THE FUTURE

As mentioned above, the number and variety of professional masters programs are increasing. As these programs become recognized, respected degree options, KGI's challenge will be to maintain the uniqueness of its programs.

Keck Graduate Institute was founded in 1997 with the narrowly focused missions of education and research dedicated to translating discoveries in the life sciences into beneficial new products and processes. That narrow focus has been a great advantage during KGI's start-up phase, and at present remains unaltered. This focus is reflected in the mission statement for the MBS degree:

> *"The professional Master of Bioscience program educates a cadre of technically competent professionals for the bioscience industry who can oversee development of useful new technologies, products, processes and services from basic life sciences research, and address the business and ethical challenges of management."*

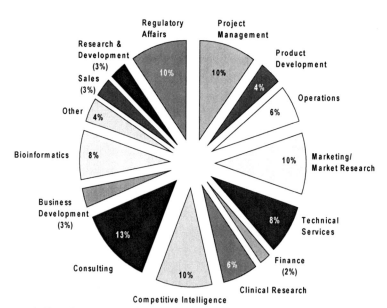

Figure 4: Employment of KGI graduates by job function

As we progress to a more mature stage of institutional development, we continue to develop our curriculum so that it remains true to the mission and at the cutting edge, while still serving the changing needs of our students and the bioscience industry. We strive to improve the integration of technology and management and provide more relevant industry experience. We rely on both our industry network and our alumni in the workforce for advice along these lines since a thorough understanding of workforce dynamics and needs is crucial to the long term success and sustainability of our enterprise.

Carolina Student Biotechnology Network

Jessica R. McCann, Shantanu Sharma, Brant Hamel, and Justin T. Brown

Jessica R. McCann is Vice President of the Carolina Student Biotechnology Network and a Ph.D. candidate in the Department of Microbiology and Immunology at the University of North Carolina at Chapel Hill. Jessica can be contacted at *jmccann@med.unc.edu*.

Shantanu Sharma is External Communications Officer of the Carolina Student Biotechnology Network and a Ph.D. candidate in the Department of Biochemistry and Biophysics at the University of North Carolina at Chapel Hill. Shantanu can be contacted at *shantanu@unc.edu*.

Brant Hamel is current president of the Carolina Student Biotechnology Network and a Ph.D. candidate in the Department of Biochemistry and Biophysics at UNC. Brant can be contacted at *brant@unc.edu*.

Justin T. Brown is Founding President of the Carolina Student Biotechnology Network and a Ph.D. candidate in the Department of Pharmacology at the University of North Carolina at Chapel Hill. Justin is also a co-founder of drug development start-up Effipharma, Inc. Justin can be contacted at *jtbrown12@gmail.com*.

The Carolina Student Biotechnology Network (CSBN) is a non-profit, student run organization that fosters career development, networking, and entrepreneurship. CSBN was created with the goal of providing information about career opportunities outside of academia, and grew out of a desire to take advantage of the University of North Carolina's proximity to Research Triangle Park (RTP), home to hundreds of life science companies that range from start-ups to multinational corporations. The organization has evolved and expanded the initial mission to include entrepreneurship and a greater focus on networking. We address our mission through two primary avenues: 1) by providing a forum for exploration of careers outside of academic science, and 2) by creating opportunities for members to acquire valuable skills and experience outside of their respective graduate programs.

CSBN has an interdisciplinary membership comprised of students and postdoctoral fellows from a breadth of graduate and professional programs such as biomedical sciences, chemistry, business, law and public health. The majority of our members, however, are drawn from research departments within the medical school and chemistry department. Membership is open to students at all regional universities, as well as industry professionals.

ORGANIZATIONAL STRUCTURE

CSBN is governed by a board of directors consisting of nine members, five students/postdoctoral fellows and four professionals from RTP or the university. General management is executed by the five students/postdocs that constitute the executive committee. The remaining board members serve to provide valuable input and important contacts but are largely uninvolved in routine tasks.

A downside to the organizational structure of CSBN is that graduate students, who are also working on their dissertation research full-time, are entirely responsible for operating CSBN. Delegation of tasks equally among executive members is critical to make such responsibility manageable in addition to graduate studies. Program directors are recruited to focus on specific areas such as communications and career development. UNC medical journalism students have volunteered to serve as the director of communications and also compose feature articles for our monthly *CSBN Access* newsletter. To further decrease workload, we solicit student representatives in each of the departments within the schools of medicine, public health, and in the business school to help advertise events and promote CSBN.

ACTIVITIES

CSBN hosts a variety of events to meet organizational objectives. These events include a seminar series, industry luncheons, panel discussions, and networking sessions. Monthly seminars are given by leaders in an assortment of fields such as biotechnology, patent law, public policy, venture capital and education. The seminars are formatted to encourage discussion and interaction between the speaker and the audience. Recent seminar speakers have included:

- Dr. John Plachetka, President, CEO and Chairman of the Board of POZEN Pharmaceuticals
- Dr. Yali Friedman, Chief Science Officer of New Economy Strategies and author of *Building Biotechnology: Starting, managing and understanding biotechnology companies.*
- Lesa Mitchell, Vice President of Advancing Innovation, Ewing Marion Kauffman Foundation
- Hannah Kettler, Program Officer, Bill & Melinda Gates Foundation
- Dr. Cynthia Robbins-Roth, entrepreneur and author of *Alternative Careers in Science: Leaving the Ivory Tower*
- Dr. Satinder Sethi, Executive Vice President of the Science and Engineering group at RTI International

Industry luncheons are hosted throughout the year to allow a small group of 8-12 members to interact with leaders in the local life science industry. The luncheons facilitate informal discussion and offer terrific networking opportunities in an intimate, relaxed setting. Participants have included companies such as GlaxoSmithKline, SAS, BD Biosciences, and Inspire Pharmaceuticals. The industry luncheons consistently rank as one of our most popular events.

Panel sessions provide an opportunity for in-depth examination of topics relevant to CSBN's mission. These sessions typically last 2-3 hours and include a break for refreshments and socializing halfway through the event. Past sessions have focused on subjects such as careers in teaching, and the technology commercialization process.

In addition to the networking opportunities presented by the activities discussed above, CSBN hosts more organized networking events at picnics on campus and in local bars and restaurants. These networking sessions often follow a seminar and allow members to not only meet speakers, but also to interact with students from other departments and institutions.

An important objective in the formation of CSBN was to increase student exposure to the plethora of events that occur in the Triangle region. Local organizations provide abundant opportunities for CSBN members to acquire diverse skills and network with local professionals. Trainees were failing to take advantage of the terrific resources that exist in Research Triangle Park, and it was evident that this failure was largely due to a lack of exposure to such resources. To ameliorate this situation we publicize relevant regional events on the CSBN website as well as in the *CSBN Access* newsletter. CSBN not only promotes events hosted by local organizations, but we have also formed relationships with several organizations such as the Council for Entrepreneurial Development, North Carolina Biotechnology Center, UNC Office of Technology Development, the Kenan Institute at UNC Kenan-Flagler School of Business, and the NC Medical Device Organization. Such associations encourage CSBN members to participate in events they may otherwise have not attended, and also allow CSBN to influence programmatic decisions made by regional organizations.

FUND-RAISING

Capital is essential to the success of CSBN. The executive committee of the organization voted against soliciting dues from members. Instead, we chose to approach institutional units and local corporations about sponsorship. Initial funding for CSBN was provided by the Carolina Entrepreneurial Initiative, an organization dedicated to spreading entrepreneurship across the UNC campus.

Potential sponsors are briefed on both the early successes and future plans of CSBN. There are three areas that we believe must be addressed when conversing with potential sponsors. First, it is necessary to clearly elucidate the need for CSBN on campus (as well as Research Triangle Park) and demonstrate how the organization will meet such needs. Second, we convey our commitment to the CSBN and emphasize that a talented, motivated team has been assembled to lead the organization. Third, it is essential to demonstrate a vision for the future and ensure sustainability beyond the university tenure of the current executive members. The last point has been especially important, as potential sponsors have expressed concern about sustainability beyond the graduation date of our founding president. Companies investing time and money into the organization want to be assured that lasting, productive relationships will be formed.

Sponsors are granted benefits dependent on the level of their donation. Benefits include recognition at events, company logo on flyers, in the *CSBN Access* newsletter and on the CSBN homepage, and the ability to send job openings to CSBN members. Official sponsors of CSBN have included: the Carolina Entrepreneurial Initiative, JMP Genomics (a business unit of SAS), Inspire Pharmaceuticals, the law firm of Wyrick, Robbins, Yates & Ponton, and Pappas Ventures.

COMMUNICATION AND COMMUNITY

CSBN utilizes a detailed and comprehensive website[1], along with frequent emails to the CSBN list-serv subscription base, to communicate with members. The website features information on CSBN events, job postings, frequent biotech news updates focusing on industry in the Research Triangle, and a blog maintained by the CSBN executive officers.

A printable copy of our monthly e-newsletter, *CSBN Access*, is sent to all members. The newsletter typically features interviews with leaders in the local life science industry, details concern-

---

1   *www.carolinabiotech.org*

ing CSBN news and events, and entrepreneurship events in the Triangle area. The newsletter also promotes relevant events hosted by other organizations in the Triangle area, including the three major academic institutions in the region: UNC-Chapel Hill, Duke University, and North Carolina State University. Cost-free distribution is a significant attribute of the monthly e-newsletters.

An objective of CSBN is to enhance career development for young scientists, and to this end our website features useful articles on topics such as writing effective resume/curriculum vitae and tips for interviewing. Career opportunities in the RTP and other biotechnology hubs around the country are listed on our career webpage, and can be discussed along with other topics in the CSBN online discussion forum at *www.carolinabiotech.org* forum. Presentations given by past seminar speakers are archived on the CSBN website along with speaker biographies.

We have recently created a CSBN blog[2] that is maintained by the executive committee and other members. The blog includes posts on events and articles relevant to CSBN's mission, as well as entries composed by members of the executive committee that address specific topics of interest. The blog provides another important mode for communicating with our members and creating an active, informed community.

Internet-based communication is an integral part of CSBN. A strong web presence has allowed an expansion in the scope of services offered by CSBN and has increased our impact in the Triangle area biotechnology community.

IMPACT

The Carolina Student Biotechnology Network has had a clear impact on campus as well as in the local life science community. While we are working on devising metrics to provide more quantitative measures of our effect, the qualitative impact of CSBN is palpable. We have successfully met our objective of providing

---

2   *www.carolinabiotech.org/blog*

opportunities for members to explore careers outside of academic science. We have facilitated the interaction of individuals from disparate fields resulting in fruitful exchange of ideas. There has been a noticeable increase in career-focused events hosted by other campus organizations and training programs, whereas prior to the existence of CSBN such events were exceedingly rare. CSBN provided the impetus for the creation of a scientific track within the Graduate Certificate Program in Entrepreneurship funded by the Carolina Entrepreneurial Initiative; in fact, CSBN's leadership has assisted in the development of this program. The certificate program will allow graduate students in the sciences to take courses in the Kenan-Flagler School of Business and will result in a capstone project that links the work done in the program to the student's scientific field.

CSBN members have been active participants in the technology commercialization process. Members have assisted with the creation Knowble Inc., started by a UNC undergraduate. Members have also worked with technology development associates within UNC's Office of Technology Development on projects dealing with financial valuations for UNC technology, and have assisted with the formation of start-up companies based on University technology. CSBN members have submitted proposals to UNC's business plan competition and participated in the Launching the Venture program, designed to help participants launch viable ventures.

FUTURE

CSBN will continue to offer popular programs previously discussed while creating innovative new programs that benefit our membership. Future directions will focus on increasing access to entrepreneurial opportunities, touring local industry facilities, creating mentoring opportunities, and organizing networking events focused on specific areas of interest. A major priority is to continue to facilitate the interaction of students with similar interests from disparate fields, as well as increasing the participation of industry professionals in CSBN. We will continue to work with established RTP organizations to leverage their resources to benefit our members. For example, we intend to augment our entrepreneurial offerings by partnering with organizations like the Council for Entrepreneurial Development and Carolina Entrepreneurial Initiative. We will strive to continue to meet our broad objectives of facilitating the development of skills necessary for future success, as well as contributing to the growth of the regional life science community.

About the Editor

Yali Friedman is author of *Building Biotechnology*, a widely used course text in biotechnology programs. He regularly guest lectures for the Johns Hopkins MS/MBA program in biotechnology, teaching classes on the business of biotechnology, and has published papers on diverse topics such as strategies to cope with a lack of management talent and capital when developing companies outside of established hubs, entrepreneurship in biotechnology, and new paradigms in technology-based economic development.

Yali is also managing editor of the *Journal of Commercial Biotechnology* and serves on the science advisory board of Chakra Biotech and the editorial advisory boards of the *Biotechnology Journal* and *Open Biotechnology Journal*.

Yali has a long history in biotechnology media, having created a Forbes "Best of the Web"-rated web site on the biotechnology industry for a NY Times company and managed it for many years. His other projects include the Student Guide to DNA Based Computers, sponsored by FUJI Television, BiotechBlog.com, and DrugPatentWatch.com, a pharmaceutical industry competitive intelligence service.

Yali can be contacted at *info@thinkbiotech.com*.

Related titles from Logos Press

BUILDING BIOTECHNOLOGY

27 Figures, 14 Tables, 35 Case Studies
Second Edition
ISBN: 978-09734676-3-5

BEST PRACTICES IN BIOTECHNOLOGY
BUSINESS DEVELOPMENT

Eleven chapters from biotechnology industry experts
ISBN: 978-09734676-0-4

3m

Printed in the United States
113315LV00004B/272/P

9 780973 467673